A SENSE OF
URGENCY

A SENSE OF URGENCY

A HYDROGEN PLAN FOR ENERGY INDEPENDENCE, ECONOMIC PROSPERITY, AND CLEAN AIR

DON BONGAARDS

This book was printed in the United States of America.

To order additional copies of this book, contact:
Xlibris Corporation
1-888-795-4274
www.Xlibris.com
Orders@Xlibris.com
52388

CONTENTS

To my grandsons Britt, Stewart, Johnathan, and Evan. May the ideas presented in this book help to make a better world for them.

About the Cover

The front cover of this book illustrates the Magenn floating air rotor. The air rotor is a lighter-than-air rotating blimp that captures the power of the wind to generate electricity and transmit it to ground level by means of a tether cable.

Acknowledgments

I would like to acknowledge the (1) Phoenix Project Foundation for many of the facts and ideas cited in this book and (2) Magenn Power Inc. for their amazing floating air rotor design.

(1) Pheonix Project Foundation, 5093 Mountain Gate Circle, Lakeside, AZ 85929 (E-mail: *hb@Pheonixprojectfoundation.US*—Web: *www.Pheonixproject.net*).
(2) Magenn Power Incorporated, Windmill Center, 105 Schneider Road, Kanata, Ontario, Canada, K2K1Y3 (E-mail: *info@magenn.com*—Web: *www.magenn.com*)

Introduction

A Sense of Urgency is a book unlike any other book ever written. It is a book about a vision of the future. Thomas Jefferson envisioned a great country based upon freedom—and it came to pass. Walt Disney envisioned a children's paradise—and it came to pass. Henry Ford envisioned a mass-produced automobile that almost everyone could afford—and it came to pass. I will propose innovative ideas (things that never were), in this book, which have the potential of providing a better life for almost everyone in the world—and whether it comes to pass or not may depend on you.

Originally, this book was intended to focus upon hydrogen as the solution to the world's *transportation* fuel crisis. However, as my research progressed, it became clear that my proposed hydrogen technology could also be used to help solve other problems facing the world—such as the generation of electricity, the increased consumption of nonrenewable resources, worldwide poverty, and global warming (if the cause is man-made).

I will begin this book by describing the *transportation* fuel crisis and how it can be solved by using hydrogen—made from wind and water. This is important for you—the reader—to understand first, since the proposed method of making, storing, transporting, and using hydrogen is a necessary first step in your understanding of how spin-off technologies can provide a better and safer world for almost everyone.

Did you know that hydrogen—the most abundant element in the universe—can be used as a fuel to power your current automobile, increase your gas mileage by 25 percent, and produce pollution-free water vapor from its exhaust pipe? Did you know that if hydrogen is used in a hydrogen fuel cell-powered automobile, it can double your gas mileage? Did you know that hydrogen can be made from water using electricity provided by the wind? And did you know that hydrogen fuel can be produced and delivered to your vehicle fuel tank for almost the same price as gasoline and can be safer to use?

Several years ago, I became concerned about why the United States Government was not emphasizing hydrogen fuel as its primary near-term transportation fuel focus. In the face of exponentially increasing worldwide oil demand and an increased dependence on foreign oil, it seemed logical to aggressively start developing a hydrogen fuel infrastructure.

As a result of preliminary research, I discovered that hydrogen fuel was being put on the "back burner" primarily because it was believed that breakthrough technologies were needed to produce, store, distribute, and use hydrogen in a safe and economical way. The words "Hydrogen is decades away" came through loud and clear. However, further research revealed that breakthrough technologies were not required, safety issues were not as much of a concern as some people believe, and that the equipment cost, to make hydrogen from wind and water, could be addressed through mass production.

Every person who uses fuel, for their transportation needs, should know how serious the current fuel situation really is. Very few people are aware of the term *peak oil* and its implication in their lives. They are also probably unaware that worldwide consumption of oil is increasing at a 2 percent annual growth rate and that energy conservation and improved efficiency—while these are good and right things to do—will only help to keep the exponential growth rate from becoming 3 percent or more. It should be made clear to these fuel users that oil sands, oil shale, synthetic fuels from coal, ethanol, and almost all other energy conservation and production methods are not long-term solutions to the *transportation fuel crisis* and that this point can be proven with simple arithmetic and logic. This does not mean to say that I am not in favor of exploiting all of these things as a short-term means of weaning the United States off imported oil and as a bridge to hydrogen.

Natural gas, can be used as a transportation fuel, and can also be a bridge to hydrogen. However, since this approach will take time and require significant vehicle and infrastructure changes, it would make more sense to leapfrog directly to hydrogen. Considering the fact Russia and the Middle East have 73 percent of the world's proven natural gas reserves, and the

United States has only 5 percent, the interim use of natural gas may also have economic and national security implications.

As T. Boone Pickens says, in his many recent radio and television ads, the United States is spending over $700 billion per year on imported oil, and something needs to be done now to correct this problem. His plan is to use wind power to reduce our natural gas consumption for electric power plants, and to use domestically produced natural gas as a transportation fuel. The Pickens plan sounds good at first, but because of the time required to make a significant impact—even in an all out Apollo Moon-type effort—there are better ways to accomplish the same thing. By the way, don't be fooled by the claim that the world has ample natural gas resources. At the current 2 percent annual consumption rate, for natural gas, the run out time is a lot closer than you may have been led to believe. I will explain more about this in the book.

My hydrogen from wind and water plan—described in this book—requires more time to make a significant impact, but a shorter term start-up plan could be implemented. This shorter term plan parallels the Pickens plan-time-wise-by making hydrogen from coal and/or natural gas. The added gas mileage, resulting from using hydrogen, will more than make up for the added cost of the conversion to hydrogen step. My plan also calls for increased drilling for domestic oil and a major increase in the production of synthetic fuel from coal. For those who are concerned about carbon dioxide emissions, the coal to synthetic fuel program can include carbon dioxide sequestration and still result in the oil equivalent of less than $80 per barrel.

It is my belief that an all out national effort toward the Pickens plan, or my hydrogen plan, will have the immediate effect of stabilizing or lowering oil and gasoline prices. This is because oil speculators will see the near future price of oil dropping, rather than increasing. Because of the vehicle and infrastructure changes required for both the natural gas and hydrogen portion of each plan, wouldn't it make more sense to implement these changes once with hydrogen rather than twice?

I believe that there is a general lack of knowledge and perspective on the part of many people who should be better informed on the subject of transportation fuels and energy in general. I find this lack of knowledge and perspective to be very disturbing. And because of my concern, I have written this book. After reading the book, it should become clear, to the reader, that hydrogen fuel is the *only* sensible solution to our near-term and long-term transportation fuel needs.

After reading this book, you will learn about energy alternatives in a way that is simplified and understandable. With this knowledge, you will be able to speak intelligently about the subject with your relatives, friends, and neighbors. You will also be able to distinguish energy facts from energy fiction when you hear the government, the media, or other people speak on the subject. Armed with this knowledge, you can also have a profound influence on the future of the United States and the world.

The most important thing to know is that the transportation fuel problem can be solved economically with hydrogen, made from wind and water, and that the wind energy by-product of this effort, along with the development of breeder fission nuclear and/or fusion nuclear, will eventually solve all of our energy problems—not just the *transportation* fuel problem. To accomplish the goals that I will set forth, we will need a program with the same *"sense of urgency"* as the Manhattan Project or the Apollo Moon Program to develop a hydrogen fuel infrastructure in a time frame that will avert economic chaos. To help achieve these goals, a new and more economical wind power technology will be introduced.

This urgent, all-out program will achieve the equipment pricing that will allow hydrogen to be economically produced and delivered. As you will see, in order to produce enough hydrogen to replace gasoline, diesel, and aviation fuels, the equipment numbers are staggering. For instance, ninety-five thousand water-to-hydrogen generators, at $10 million each, are needed. Because of these sizeable numbers, large mass production factories, which are about 70 percent vertically integrated and almost 100 percent robotic, can be justified. In this book, I will introduce an advanced mass production method, based upon these large numbers, that will make product costs even lower than current practice. Just think of the value you are getting for your money when you purchase a $25,000 mass-produced automobile. Now imagine that twenty times as many of that same automobile were produced—unchanged—over a period of ten years, and the price was not encumbered by noncontributing expenses. In this case, it is conceivable that the price could be reduced by as much as 50 percent and still produce a substantial profit.

When manufacturers see the massive numbers required to achieve our hydrogen fuel objectives, they will see the profit potential, even when challenged to reduce their selling price by as much as 70 percent. Penalties and incentives for reaching, or not reaching, their price goals would introduce a level of creative thinking that we cannot imagine at this time. The result of the manufacturing technology that will potentially be developed may set the stage for a new way of producing manufactured goods and making manufactured prices less expensive for everyone. Read this book and see why.

Also, see why it is necessary to have *"a sense of urgency"* commitment, and focus, along the lines of the Apollo Moon Project, to make it happen. It will not happen on its own, as many people believe; instead, it will require presidential leadership and a national commitment. However, when this commitment is made, the result could very well lead to energy independence, economic prosperity, and cleaner air.

Chapter 1
The Way Things Are

The future ain't what it used to be

—Yogi Berra

1.1 Where We Have Been

Yes—but what about air pollution, and depletion of our non-renewable resources!

We sure have made a lot of technological progress in the past 100 years.

I currently have four grandchildren ages twelve, nine, five, and one. When the five, and one, year old become six years of age, they will be told a bedtime story about great-grannie as their brothers had been told before.

Great-grannie was descended from the Reverend Ralph Wheelock, who arrived in Massachusetts in 1637, where he helped to establish several towns around the Boston area. One of Ralph's descendants, Eliaser Wheelock, founded Dartmouth College, and another descendant, Able Wheelock, was a loyalist to England during the revolutionary war and moved his family to Nova Scotia, Canada, where he was given two thousand acres of land by the King of England.

Great-grannie's grandfather Elarcum Wheelock, a descendant of Able, had inherited one hundred acres of Able's original two thousand acres of land. The land was eventually inherited by great-grannie's father—Maynard Wheelock.

Great-grannie was born in 1908 and was the eldest of four sisters and one brother. The farmhouse, where great-grannie grew up, had no running water or electricity, and the bathroom consisted of an outhouse and chamber pots. Water was obtained from a well, and kerosene lanterns were used for light. The farmhouse had a wood burning stove, and chopped wood was continuously fed to the stove all year long. The staple food was beans, which were continually replenished from the previous year's crop and kept heated in the stove. The farm had a twenty-acre garden that supplied food throughout the year through canning of fruits and vegetables at harvest time and storing them in a cellar. Milk was supplied by cows and eggs by hens. In the winter months, meat was supplied by deer that had been shot and kept frozen in the barn since there was no refrigerator. In the summer months, a chicken or pig was slaughtered. Some of the pork was salted and added to the beans. They grew wheat on the farm and had the local gristmill make flour. To pay the gristmill, it was agreed that the mill operator could keep and sell half of the flour. There was a country store that also bartered with the local farmers. For instance, sugar was exchanged for eggs, and molasses was exchanged for milk. Grannie's father cut hay with a sickle, and his primary mode of transportation was a horse and buggy. There was one church nearby that also served as a one-room schoolhouse, where children from all grade levels were taught by one teacher.

When great-grannie was born, the world population was about one billion people, and oil didn't have much value. Electricity, automobiles, and airplanes had just been invented and were not in widespread use. Televisions and electric dishwashers were not yet invented. Life was simpler then, and consumption of the earth's natural resources was very limited. However, when great-grannie's third grandchild—the author of this book—was born in 1941, only thirty-three years later, the world population was closing in on three billion people, and the Industrial Revolution was well underway. At that time, many people owned cars, mainly because of Henry Ford's assembly line manufacturing system that made them affordable to almost everyone. While people in China, India, Africa, and other underdeveloped countries, which comprised about 80 percent of the total world population, did not participate in the Industrial Revolution.

Today the world population is about 6.6 billion people, yet only about 30 percent are consuming oil and other nonrenewable resources. That percentage is growing rapidly as China, India, and other highly populated countries participate in the modern industrial age. Thirty percent of 6.6 billion is much more than 30 percent of three billion in 1950. Additionally, it is common today to own two or three cars per family, rather than in the '40s and '50s, when it was common to own only one car per family. The trend, in the United States, has been for people to live in the suburbs and drive a long distance to get to their place of employment.

Predicting the future can be difficult, but if you see a trend, it is probably a good bet that the trend will continue. Take the case of the man who, in the middle of the nineteenth century, reliably predicted when a snowstorm was going to hit his hometown. Everyone was amazed at how accurate he was, but he wouldn't reveal his secret until very late in his life. His secret was that before a snowstorm hit his town, he would observe that the train that passed by his house had snow on its roof, and the wind was blowing in the direction that the train was traveling.

Just like the snow on the roof of the train, the trend of an increasing demand for the world's oil supply is pretty clear, and the resulting effect is not difficult to predict. In the case of the forecasted snowstorm, nothing much could be done to keep it from happening; however, with the threat of depleting the world's oil supply, there is something that can, and should, be done.

The "what should be done" proposed in this book is to have an all-out effort to shift to a hydrogen fuel economy as fast as possible—a Manhattan-type Project. There are other proposals on the table, but as will be shown, these proposals are not real long-term solutions. They include conservation, fuel efficiency, ethanol, oil from oil sands and oil shale, synthetic fuels from coal, natural gas, methane hydrates, fission nuclear power, fuel taxes, and fuel rationing. However, when compared to the hydrogen fuel solution—to be described herein—they should be used only as transitional methods.

As you will see in the illustrated arithmetic in subchapter 3.1, the resulting price of liquefied hydrogen fuel—made from wind and water—at the gas station pump, is estimated to be about $4.63 per equivalent gallon of gasoline. Also, because hydrogen fuel is much more efficient to use than gasoline—more miles per gallon—there will be a substantial savings per fuel fill-up.

One of the main reasons why hydrogen fuel has taken a backseat to the above-mentioned methods is because of paradigms. A paradigm is seeing things a certain way because you expect them to be that way. Paradigms have a way of causing people to "back off" because they don't want to go against mainstream thinking. It's "cooler" to advocate conservation, fuel efficiency, and ethanol than it is to advocate hydrogen fuel. Unfortunately, hydrogen fuel has become associated with environmental extremists—who, in this case, are right.

There are those who say that hydrogen fuel is either too expensive or too dangerous because of what they have been told or what they perceive from incidents like the *Hindenburg*. Then there are those who say that hydrogen fuel has too many technical issues—like storage—that need a breakthrough technology to be resolved and that it will take decades to resolve these issues. Also, there are people who fear change and want to keep the status quo or who have special economic interests. In addition, there are those who don't even know that hydrogen is a component of water and that it can be used as a replacement for gasoline.

How did this antihydrogen paradigm occur? Have we been sold a "bill of goods" from those who are making money from oil or who have a vested interest in the continuation of oil? After reading this book, you can be the judge.

What will it be like when my four grandchildren, ages one to twelve, become fifty-nine to seventy-one years old? Will they say to *their* grandchildren that we were the greatest generation—as we say about our grandparents—or will they tell them that our generation was shortsighted, greedy, and foolhardy? That we didn't act soon enough on many issues, and as a result, we left them in an overpopulated world with a regressive economy?

If you look at current population forecasts, it is anticipated that the world population will grow from today's 6.6 billion to over 9.2 billion by the year 2050. At that time, will we be using our farmland to make ethanol, or will we be using it to help feed the world? Will we be polluting our atmosphere at an ever-increasing rate by burning gasoline and other carbon-based fuels, or will we be using clean burning hydrogen fuel, supplemented by fusion and/or breeder fission nuclear electric power? Will our continued rate of consuming hydrocarbon fuels actually cause the predicted effects of global warming? Will our consumption of the earth's nonrenewable resources be such that we continue to bury the remains of manufactured products in landfills, or will we emphasize engineering designs of products that can be returned to factories for disassembly and recycling?

Whatever happens, I hope that my grandchildren will tell their grandchildren that Grampa wrote a book that described how and why we needed to shift to a hydrogen fuel economy before it became too late.

1.2 Things You Ought to Know

Worldwide poverty results in hunger, disease, and despair. It also contributes to a rapidly increasing world population and may be the root cause of war. As the world population becomes more industrialized, more of the earth's nonrenewable resources are going to be consumed *with energy-producing resources such as oil and natural gas currently being the biggest nonrenewable resource depletion problem.*

My vision is that by solving the current *transportation* fuel crisis, and the energy crisis in general, a technology platform will be set for solutions to worldwide poverty. Because solving the transportation fuel crisis is—as will be shown—fundamental to how the other problems can be solved; the majority of this book will deal with this issue. To begin, we need to first understand what the transportation fuel crisis is, what is currently being done about it, and what should be done.

Did You Know That

The world's population is expected to grow from 6.6 billion people now to about 9.2 billion by the year 2050.

Thirty-five percent of the current world's population lives without electricity (2.28 billion people).

Seventy percent of the current world's population does not use gasoline, diesel, or aviation fuels for transportation (4.55 billion people).

Half of the world's current 6.6 billion people live on less than the equivalent of one U.S. dollar per day.

Since 1945, 4.6 million square miles of the world's farmland has lost much of its productivity—an area about the size of India and China combined.

Farmers are abandoning about twenty-seven thousand square miles of farmland each year because the soils have become degraded, and water is becoming scarce.

The world's fresh groundwater is being consumed about ten times faster than it is being replenished.

Worldwide disease and hunger is rapidly increasing due to the increasing population, poverty, and degradation of farmland.

The world is currently consuming transportation fuels at an increase of more than 2 percent annually—mainly because of the rapid industrialization of China and India.

At the 2 percent annual transportation fuel consumption increase, the 2050 fuel consumption will increase from today's 1.89 billion gallons per day to over 6.06 billion gallons per day.

At the current 2 percent annual transportation fuel consumption increase, the currently known world oil reserves will be depleted in about twenty years.

If *all* of the world's cropland were used to make ethanol, it would equal about 1.5 billion gallons of equivalent gasoline per day—assuming that only half of the ethanol energy is used in the process of making it.

If currently known world reserves of oil, oil sands, coal, and oil shale could be produced fast enough, and were used *only* for transportation fuels, the world would run out of these resources by the year 2067.

The world's natural gas consumption is also increasing at a 2 percent annual growth rate, and will be depleted well before the end of the 21st century if unproven reserves become available. Note: 73 percent of proven natural gas reserves are located in Russia and the Middle East.

The exponentially increasing use of fossil fuels will significantly increase worldwide health problems due to air pollution and accelerate the levels of atmospheric carbon dioxide that *may* result in the potential effects of global warming.

If all of the world's electric power were derived from currently designed fission nuclear power plants, that the uranium fuel would be rapidly depleted.

There isn't enough hydroelectric or *conventional* geothermal sites available in the world to sustain even a small fraction of *current* world energy consumption.

Current conventional fossil-fired and nuclear power plants produce electricity for about half the cost of current wind-generated electricity and a third less than the cost of current solar-produced electricity.

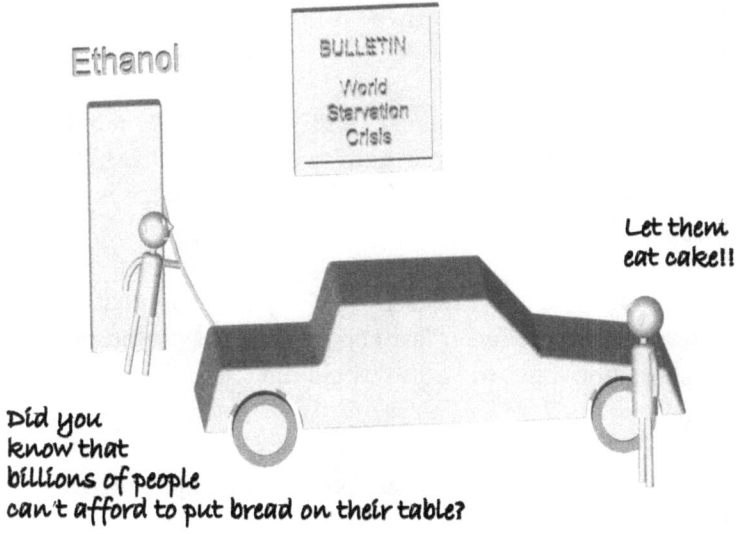

While resolution of the above problems may seem insurmountable, I will propose innovative solutions, in this book, that will not only solve these problems, but will do it in a manner that will significantly increase the world's wealth. However, the most innovative solution was a discovery of how to make cheap and abundant renewable electric energy.

Almost everyone knows that the wind and sun provide an unlimited source of energy. The only problem is that capturing this energy turns out to be expensive and intermittent relative to burning fossil or nuclear fuels. If it were less expensive and nonintermittent, we would all be using it today. Solving this problem is the "holy grail" that many engineers and scientists are trying to find.

While investigating new ideas to help solve the current transportation fuel crisis, I discovered a fatal flaw in the way that the world is currently trying to provide low-cost renewable energy from the wind. The fatal flaw was that propeller-and-pedestal wind turbines had inherent design deficiencies that will probably keep them from ever becoming widespread or competitive with current conventional or nuclear power plants—even with high-volume mass production. High-volume mass production of solar-produced energy will help bring costs down from present levels, but at present, wind-generated electricity is twice as expensive to produce as fuel-burning power plants, and solar-generated electric power is three times as expensive—and this is more than a tenfold improvement over what it has been in the recent past. Obviously if currently proposed wind and solar solutions were competitive, free market forces would take over, and government subsidies would not be necessary. Instead of 1 percent of the world's electric power being produced from the wind and sun, it would be close to 100 percent.

1.3 Thinking Outside of the Box

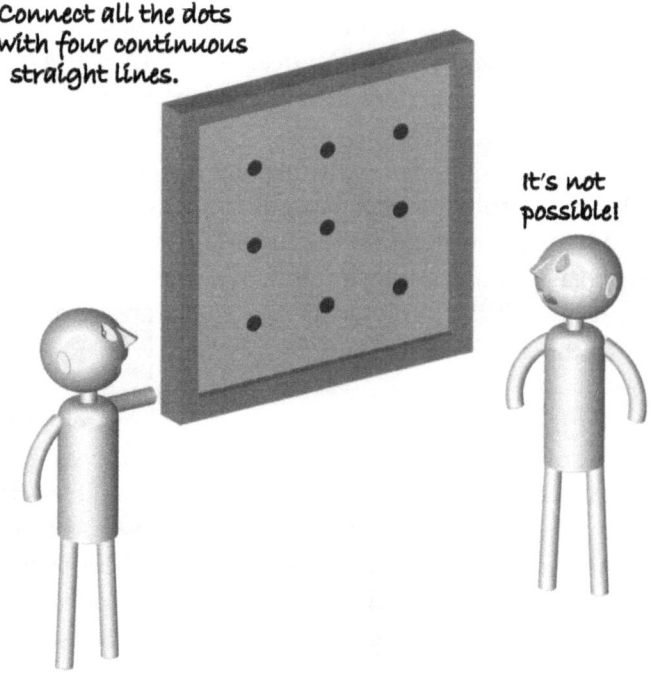

To prepare the reader for what will be presented in this book, it is important to keep an open mind. Sometimes we have preconceived ideas and worldviews—paradigms—that keep us from seeing what the facts are telling us. And even if you don't agree with what will be presented in this book, it would be prudent to at least know, and understand, the other side of the story.

Have you ever been given the "out of the box" presentation? It is a favorite of many lecturers who want to make a point about expanding your thinking process beyond what you may believe is possible. See the illustration below:

QUESTION
Connect all of the dots with four straight lines
without lifting the pencil from the paper

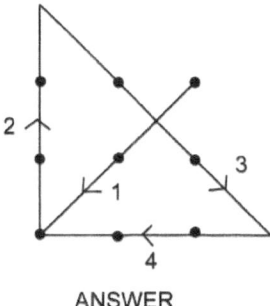

ANSWER

As you can see, there are nine dots that need to be connected with four straight lines, and the lines need to be drawn continuously. As shown in the answer, it is necessary to go "outside of the box." What this exercise illustrates is that too many times, we confine our thinking to being inside a box as if our mind had boundaries or limitations. As you read this book, I hope that you will stretch your mind beyond what you may currently think is possible.

To illustrate this further, let's take the case of a New York City planner at the turn of the nineteenth century. When asked how many people could live in the city, he responded by saying one million. His answer was based upon the idea that there was only enough space to stable horses and carriages for about one million people. Do you get the point?

When you begin to criticize the contents in this book, especially with regard to hydrogen fuel cost estimates and new ideas regarding electricity generated from wind power, remember the "out of the box" illustration and the New York City planner.

Necessity is the mother of invention, and just because you may have "boxed" in your thinking, it doesn't mean that others have. When Bill Gates, who may be the richest man in the world, worked for IBM, he was asked to develop an operating system for a new thing called a personal computer. At that time, large "mainframe" computers were IBM's bread and butter. The personal computer was thought to be a toy that would be used by a limited number of people, or something that engineers and scientists would use to access "mainframe" computers. Because of IBM's limited thinking, they required Bill to make the IBM personal computer compatible with their mainframes. As a result, it took about fifteen minutes for Bill's operating system to get the personal computer started (booted up). Bill advised IBM to remove the mainframe requirement from the operating system so that the computer could start up in about one minute—rather than fifteen. IBM's management refused because of their perception of the future. Bill's vision of the future of personal computers was quite different. And Bill was so convinced of his vision that he started his own company—Microsoft. While he was doing this, IBM almost went out of business, and you all know the rest of the story.

The next two chapters of this book will be devoted to describing the transportation fuel crisis and how it can be solved with new and innovative ideas. With the background provided in these chapters, you will be better prepared for the fourth chapter entitled "Things That Never Were." As you will see, it is the spin-off of ideas used to solve the transportation fuel crisis that can be applied to solve other problems facing the world today.

So as you will soon find out by thinking "outside of the box," we can, and should, dream of "things that never were" and say, "Why not!"

Chapter 2

The Energy Crisis

When you can measure what you are speaking about, and express it in numbers, you know something about it; but when you cannot measure it, when you cannot express it in numbers, your knowledge is of a meager and unsatisfactory kind; it may be the beginning of knowledge, but you have scarcely, in your thoughts, advanced to the stage of science.

—Lord Kelvin (1821-1907)

2.1 The Oil Problem

Currently, the world consumes over 85 million barrels of oil each day, of which 21 million barrels is consumed by the United States. Of the 21 million barrels consumed by the United States, 5.25 million barrels per day come from the Middle East. At more than $100 per barrel (up from about $42 per barrel at the beginning of 2005), with 42 gallons of oil per barrel, this amounts to about $192 billion per year *(Note: The current price of oil is much higher than $100 per barrel, but for simplicity, I will use $100 per barrel for the remainder of this book)*. If nothing is done about this situation, the Middle East portion of our oil purchases will probably grow from the present 25 percent to more than 30 percent by the year 2025.

Couple this with the concern that many in the United States believe that a terrorist attack on a large Saudi oil field could cut world production by 10 percent and quickly send oil prices soaring.

But most importantly, worldwide consumption of oil has been increasing very rapidly. From 1999 to 2003, the world oil consumption increased 5 percent. From 2003 through 2005, oil consumption increased 6 percent. This exponential increase in oil consumption was, and is, due primarily to a fast-growing middle class in China, India, and other developing nations.

To further exacerbate the situation, we are currently finding only one barrel of oil for every four barrels of oil being consumed, and existing oil wells are starting to "dry up." For instance, the Alaska pipeline production—traditionally at one million barrels of oil per day—is now declining because of peaking of the oil reservoir in Prudhoe Bay.

What this means is that we are approaching a point in time where the world oil production will "peak." In other words, the world's oil reservoirs will reach a point where their total rate of production cannot increase. It also means that the production rate will decrease with time. However, it does not mean that the oil wells have run dry; instead, they just cannot produce oil at a faster rate.

It is a well-established fact that output of individual oil reservoirs (wells) rises after discovery, reach a peak, and decline thereafter. Oil reservoirs have lifetimes typically measured in decades, and peak production occurs roughly about a decade or so after discovery. "Peaking," for an individual reservoir, is its maximum oil production rate and typically occurs after roughly half of the recoverable oil in the reservoir has been produced.

The lower forty-eight states were one of the richest, most geologically varied, and most oil-productive areas in the world up until 1970. When oil prices tripled in 1973 to 1974, and by another factor of two in 1979 to 1980, oil exploration was increased on a large scale. This exploration included three-dimensional seismic analysis, economic horizontal drilling, and dramatically improved geological understanding.

Nevertheless, oil production trended downward. In light of this experience, there is good reason to believe that an analogous situation will exist worldwide, after oil production peaks. *Higher prices and improved technology are unlikely to yield dramatically higher oil production.*

Oil reserves have been in a decline relative to consumption since 1985. This is coupled with the fact that new "super giant" oil reservoirs—mainly in the Middle East—that are the easiest to find, the most economical to develop, and the longest-lived, have not been found since 1967 and 1968. Since then, smaller reservoirs of varying sizes have been discovered in what are called "oil prone" locations.

What this indicates is that the world peak-oil production rate is probably not too far into the future. Listed in table 2.1.1 are projections from credible analysts on the subject of "peaking" of world oil production:

Table 2.1.1

Table 1. Projections of the Peaking of World Oil Production		
Projected Date	*Source of Projection*	*Background & Reference*
2006–2007	Bakhitari, A. M. S.	Iranian oil executive
	Simmons, M. R.	Investment banker
After 2007	Skrebowski, C.	Petroleum journal editor
Before 2009	Deffeyes, K. S.	Oil company geologist (ret.)
Before 2010	Goodstein, D.	Vice Provost, Cal Tech
Around 2010	Campbell, C. J.	Oil company geologist (ret.)
After 2010	World Energy Council	Nongovernmental org.
	Laherrere, J.	Oil company geologist (ret.)
2016	EIA nominal case	DOE analysis/information
After 2020	CERA	Energy consultants
2025 or later	Shell	Major oil company
No visible peak	Lynch, M. C.	Energy economist

Table 2.1.2

Oil reserves (millions of barrels as of January 1, 2002): Top 20 countries

Saudi Arabia: 261,750
Iraq: 112,500
United Arab Emirates: 97,800
Kuwait: 96,500
Iran: 89,700
Venezuela: 77,685
Russia: 48,573
Libya: 29,500
Mexico: 26,941
Nigeria: 24,000
China: 24,000

United States: 22,045
Qatar: 15,207
Norway: 9,947
Algeria: 9,200
Brazil: 8,465
Oman: 5,506
Kazakhstan: 5,417
Angola: 5,412
Indonesia: 5,000

Note: First 20 countries: 975,148; rest of the world: 56,983

Oil consumption(millions of barrels per day): Top 20 countries

United States: 19,993
Japan: 5,423
China: 4,854
Russia: 2,531
South Korea: 2,126
Brazil: 2,123
Canada: 2,048
France: 2,040
India: 2,011

Mexico: 1,932
Italy: 1,881
United Kingdom: 1,699
Spain: 1,465
Saudi Arabia: 1,415
Iran: 1,109
Indonesia: 1,063
Netherlands: 881
Australia: 879
Taiwan: 846

Note: First 20 countries: 59,134; rest of the world: 16,854

Oil production (millions of barrels per day): Top 20 countries

Saudi Arabia: 8,528
United States: 8,091
Russia: 7,014
Iran: 3,775
Mexico: 3,560
Norway: 3,408
China: 3,297
Venezuela: 3,137
Canada: 2,749
United Arab Emirates: 2,550

United Kingdom: 2,540
Iraq: 2,377
Nigeria: 2,223
Kuwait: 1,838
Brazil: 1,589
Algeria: 1,486
Libya: 1,427
Indonesia: 1,384
Oman: 964
Argentina: 825

Note: First 20 countries; 62,762; rest of the world: 12,464

Source: http://www.thirdworldtraveler.com/oil reserve

The extensive exploration and drilling for oil and gas has provided a massive worldwide database, and current geological knowledge is much more extensive than in past years. In addition, various seismic and other exploration technologies have advanced dramatically in recent years.

The fact is nobody knows how much oil supply we have left. But what we do know is that when an oil well runs dry, it happens very rapidly. And if a "super giant" oil well runs dry, the short-term consequences could be devastating to gasoline prices.

To put things further into perspective, many people quote journalists and experts who claim that we have thirty to forty years of oil reserves. These reserves are shown in table 2.1.2 (known information through 2001):

Although the above numbers are over seven years old, the world reserves have been depleted by its usage over this time period, and the current barrels per day consumption rate (usage) has increased to more than 85 million barrels per day—an almost 12 percent increase in five years.

If we assume that the above numbers are correct for the year 2001, the world oil reserves were 1,032,132 million barrels, and the world consumption rate was 75.988 million barrels per day, *but increasing at a rate of 2 percent per year*—let's do some arithmetic:

1. By the year 2005, the consumption rate is calculated to be 82.3 million barrels per day, and the remaining reserves would be 1,002,110 million barrels (note that the actual consumption rate for the year 2005 was more than 83 million barrels per day).

2. By the year 2010, the consumption rate would equal 90.8 million barrels per day, and the remaining reserves would equal 728,435 million barrels.

3. By the year 2020, the consumption rate would equal 110.7 million barrels per day, and the remaining reserves would equal 358,229 million barrels.

4. By the year 2028, all of the currently known oil reserves would be gone, and the world consumption rate at that time would equal 129.703 million barrels per day.

Some readers may be saying twenty years is a long time, and I'm not going to worry about it now. Besides, didn't we just find a huge oil reserve in the Gulf of Mexico,* and if permitted, we could drill for oil in the Arctic

---

* Three oil companies, led by Chevron Corp., have found what could be the largest oil discovery since Alaska's Prudhoe Bay a generation ago. The estimated 300-square-mile region, in the Gulf of Mexico, could hold 3 to 15 billion barrels of oil and natural gas liquids, located 28,175 feet (5.34 miles) below sea level. The Chevron "Jack 2" oil rig initially produced six thousand barrels of oil per day and is expected to increase significantly in the near future.

National Wildlife Refuge (ANWR)? OK—the above chart does not reflect this new Gulf of Mexico find or the ANWR.

The estimates for the Gulf of Mexico oil discovery are from 3 to 15 billion barrels of oil. This is an increase of 1.5 percent in world oil reserves, if the 15-billion barrel figure is realized. The oil in the ANWR may even double this figure to 3 percent. All this does is verify what was stated above by showing how difficult it is to find large amounts of oil.

However, this new oil, along with other mitigating methods, to be discussed in the next subchapter, does help to make the United States more energy independent for the short term, and I am all for it.

2.2 Mitigation

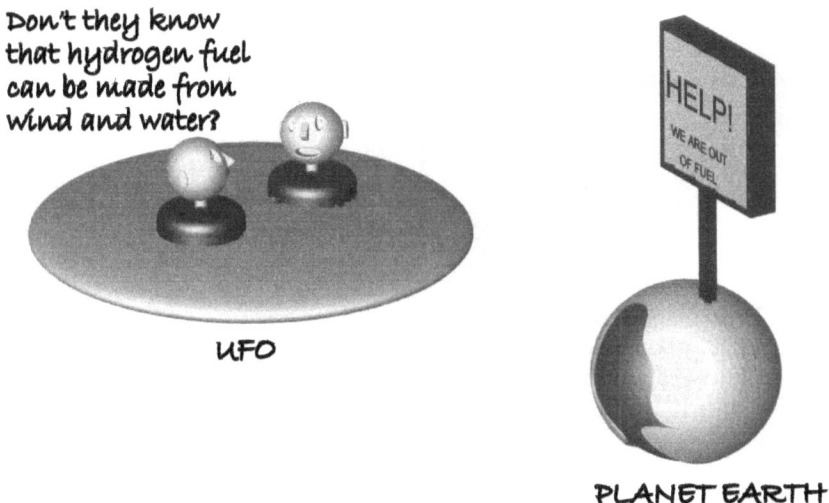

To mitigate the ominous transportation fuel crisis presented in the previous subchapter, the United States, and other countries around the world, are increasingly turning to conservation, energy efficiency, ethanol, oil sands, synthetic fuel from coal, natural gas, and oil shale. As you will see, in this subchapter, these mitigating methods, if used as long-term solutions, are like wrestling for deck chairs on the *Titanic*—they do not address the inevitable, and we will eventually hit an energy iceberg!

Conservation and Energy Efficiency

Conservation and energy efficiency—like motherhood and apple pie—can certainly help stave off the ominous fuel shortage threat, but the world population is rapidly increasing and becoming more industrialized. As a result, this mitigating method will only help to keep the exponential rate of transportation fuel increase to only about 2 percent, but it only delays the inevitable—more on this later.

Ethanol

The subject of ethanol—or biofuel—is very complex and is generally misunderstood. This is because the processes used in its formation varies significantly. Probably the best and most promising method for producing ethanol is shown below:

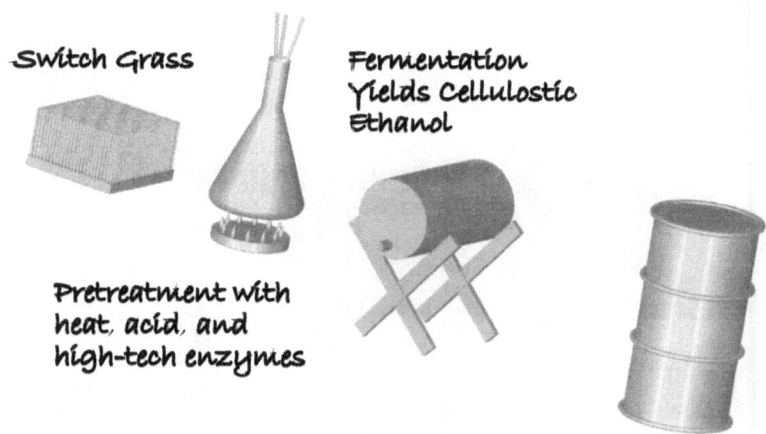

Switch Grass

Fermentation Yields Cellulostic Ethanol

Pretreatment with heat, acid, and high-tech enzymes

There are several reasons why this is the best method:

Switchgrass—deep-rooted perennial prairie grass—can have multiple crops during a year and can be grown on marginal land with relatively small amounts of water and fertilizer. The energy that is put into the process can result in an energy output of between 2 and 36. At the high end, greenhouse gas emissions—production and use—can be reduced by 91 percent.

The problem with reaching the potential high-end numbers is that the process currently does not exist, can be very expensive, and depends on growing algae rather than switchgrass. With regard to switchgrass, and other fast-growing plants, researchers are currently trying to find better, more efficient ways of breaking down the cellulostic material into sugars that can be fermented. The sugar extraction is accomplished by loosening the bond between the cellulose/hemicellulose and lignen. One possibility is to produce genetically modified microbes and enzymes from the guts of termites—nature's own cellulostic energy factories. At present, an input-to-output energy ratio of 4 is feasible but expensive.

The current—most widely used—method of producing ethanol is to convert food crops, like corn, sugar beets, soybeans, and sugarcane, since these crops do not require exotic methods to extract their sugar content. From an energy balance standpoint, corn is the worst at about 1 unit of energy input versus about 1.3 units of energy output. Sugarcane, on the other hand, can have an output-to-input ratio of about 8 to 1.

Because sugarcane is such an effective ethanol producer, Brazil, with its warm and moist climate, is producing an ever-increasing amount of sugarcane ethanol. In fact, it plans to double its cane acreage over the next decade. Thus contributing to deforestation and displacing existing agricultural areas. As a result, cattlemen are being driven deeper into frontier territory like the Amazon and biologically diverse savannas known as the *cerrado*.

In the United States, corn is the leading producer of ethanol. While at first this may not seem to make sense, keep in mind that we are trying to free ourselves from dependence on Middle East oil. As such, we are laying the groundwork for an ethanol infrastructure that should become more reasonable as the cellulostic methods are being developed. In fact, if the corn stocks could be converted into ethanol, it would increase corn's energy ratio from 1.3 to 1 to more than 2 to 1.

To be clear about how ethanol can reduce greenhouse gas emissions, it is assumed that by burning ethanol in your automobile, you are, in effect, producing carbon dioxide that would be reabsorbed back into the plant's perennial growth cycle. But this does not account for carbon dioxide that is emitted during its production process.

Although ethanol has problems that must be accommodated, such as 39 percent lower-energy content per gallon than gasoline, a potential engine corrosion problem, if exposed to water, and the potential for using hydrocarbons in its distilling process, it does help to provide a short-term solution to the fuel crisis problem.

When ethanol is used as a long-term solution, even as an energy mix with other fuel solutions, it cannot be sustained. Why? Because with the world's population increasing from 6.6 billion now to about 9.2 billion by the year 2050, we will have a food shortage problem. In fact, today, twenty-five thousand people die of hunger every day—most under the age of five.

This food shortage is due in part to agricultural land losing its productivity. For instance, since 1945 much of the agricultural productivity of the world's farmland has been lost. This lost farmland amounts to an area about the size of India and China combined. Today, farmers abandon about twenty-seven thousand square miles of farmland each year because the soils are too degraded for crops.

If the lost farmland could be restored, and cellulostic ethanol could be made from crop residues and switchgrass—rather than the food produced by the crop—ethanol may start to make sense; more on this later.

Oil Sands

The process for making crude oil from oil sand is illustrated below:

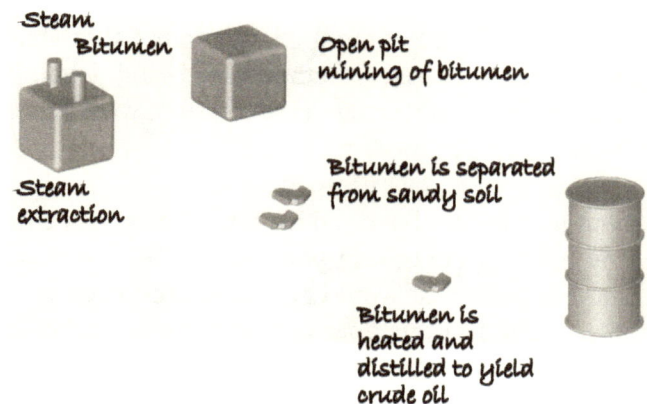

Steam
Bitumen

Open pit
mining of bitumen

Steam
extraction

Bitumen is separated
from sandy soil

Bitumen is
heated and
distilled to yield
crude oil

Oil sand is a thick sticky form of crude oil and must be extracted from a complex mix of sand, water, and clay. The heavy syrup is either drawn out with heat or mined with machinery and then heated to transform it into usable oil. About a barrel of water is used for every barrel of oil produced, and the process takes large volumes of natural gas to heat and separate the oil from the sand.

The total world's oil sand deposits equal about 485,000 million barrels of recoverable oil (http://en.wikipedia.org/wiki/oil_sands), with 75 percent being located in Canada and Venezuela.

Canada—primarily in northern Alberta—has one of the richest oil sand deposits in the world, with recoverable oil equal to 178,000 million barrels.

At this point in time, the recovery of oil from oil sands is the least expensive method of producing an alternative source of oil. And at the present time, Canada's oil production, from oil sands, is 1.2 million barrels per day and is expected to double over the next four years.

Synthetic Fuel from Coal

The process for making synthetic fuel from coal is illustrated below:

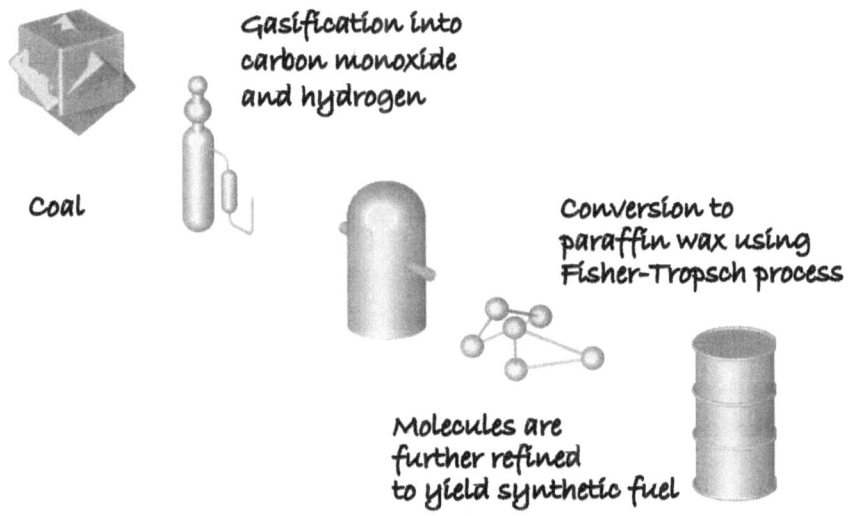

Coal

Gasification into carbon monoxide and hydrogen

Conversion to paraffin wax using Fisher-Tropsch process

Molecules are further refined to yield synthetic fuel

The method used for making synthetic fuels is called the Fisher-Tropsch (FT) process and is described by the following chemical equation:

$$CH_4 + 1/2O_2 \rightarrow 2H_2 + CO$$
$$(2n + 1)H_2 + nCO \rightarrow CnH_2n + nH_2O$$

The mixture of carbon monoxide (CO) and hydrogen (H_2) is called synthesis gas or syngas. The resulting hydrocarbon products are refined to produce the desired synthetic fuel.

This process was invented in petroleum-poor, but coal-rich Germany in the 1920s to produce liquid fuels. It was used by Germany and Japan during WWII to produce alternative fuels, and Germany's synthetic fuel production reached more than 124,000 barrels per day from twenty-five plants.

After the war, captured German scientists, recruited in Operation Paperclip, continued to work on synthetic fuels for the United States Bureau of Mines program initiated by the Synthetic Liquid Fuels Act.

Currently, two companies have commercialized their FT technology, Shell in Bintulu, Malaysia, and Sasol in South Africa. A small U.S. company, Rentech, is also producing liquid hydrocarbons as a by-product of producing fertilizer.

The FT process is an established technology and is already being applied on a large scale. However, its popularity was hampered by high capital costs, high operating costs, high operation and maintenance costs, for what was a relatively low price of crude oil. This is currently not the case at $100+ per barrel.

The one issue that has yet to be addressed in the large-scale development of synthetic fuels is the enormous primary energy use and carbon emissions inherent in the conversion of gaseous or carbon sources to a usable liquid form. Recent work by the National Renewable Energy Laboratory (NREL) indicates that full fuel-cycle greenhouse gas emissions—carbon dioxide—for coal-based synfuels are nearly twice as high as their petroleum-based equivalent. Emissions of other air pollutants are vastly increased as well. Carbon sequestration has been suggested as a strategy for capturing greenhouse gas emissions. However, while sequestration is already in limited use, the science and economics of large-scale sequestration strategies are as yet unproven—more on this later.

Some have proposed using nuclear power plants to provide the intense heat needed—2500 °F—to make synthetic gasoline from coal. If this were done, the net increase of atmospheric carbon dioxide would be zero, since it would merely replace existing consumption of oil-based fuel.

The United States Department of Energy (DOE) estimates that there are 2,002 trillion pounds of recoverable coal in the world, and that the United States, with 27 percent, has the largest share. If all of the world's coal were converted into synthetic fuel—at 50 percent efficiency—it would be the equivalent of 3,557 billion barrels of synthetic fuel—or 3.92 times the current world oil reserves after conversion to gasoline. However, a large portion of this coal is currently being used for electric power generation. And like oil, it is being consumed at an exponential rate as worldwide electric power is increasing. For instance if the current U.S. coal usage were converted into transportation fuel it would be equal to about 60 percent of the current U.S. transportation fuel usage. With this fact in mind, it should not be assumed that the if coal were used to make synthetic fuel, that it would last as long as some may think—more on this later.

Currently there are four U. S. companies constructing coal to synthetic fuel plants. Their combined output will be about 132,000 barrels per day. Projections are that they will produce synthetic fuel that is equivalent to about $57 per barrel of oil, and they plan to come on line between 2011 and 2013. This will result in 1.2 percent of the total transportation fuel consumption in the United States.

Considering the fact that the United States is currently spending about $700 billion for imported oil, and is an economic turn down, does it make sense to not vigorously pursue the coal to synthetic fuel option? Why aren't we building 500 coal to synthetic fuel plants and sequestering the carbon dioxide? Surely the added cost of sequestration will not increase the equivalent barrel of oil production cost to more than $80, especially if the plants are located on sites that are conducive to this process. Also consider what spending $700 billion in the United States would do for its economy in terms of jobs and economic growth that will have the effect of increasing tax revenue and reducing the deficit. Go figure!

Oil Shale

The process for making liquid fuel from oil shale is illustrated below:

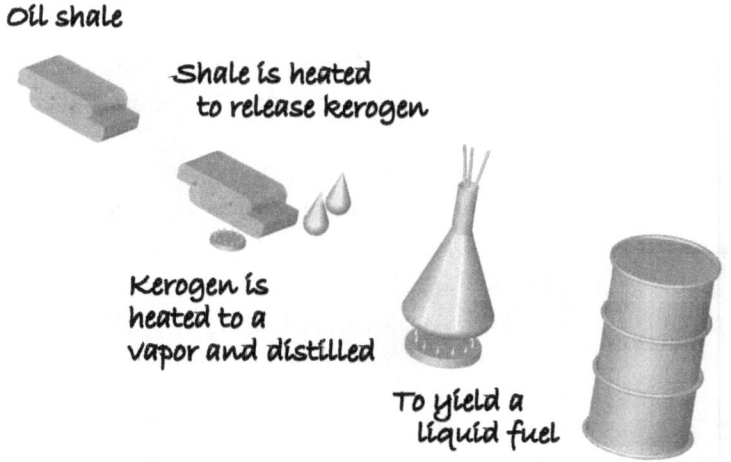

Oil shale

Shale is heated
to release kerogen

Kerogen is
heated to a
vapor and distilled

To yield a
liquid fuel

The World Energy Council (WEC) estimates that recoverable world oil shale reserves are about 620 billion barrels of oil, with 94 percent located in North America.

The United States has by far the largest oil shale deposits that are located in a 16,000 square mile area of the Green River region of Colorado, Utah, and Wyoming. A substance called kerogen is produced from oil shale, and the Green River area has reserves equal to about 560,000 million barrels.

Oil shale is buried beneath about 5,000 feet of earth and can be 11,000 feet thick. Rather than dig the largest open pit mine in the world, Shell Oil's proposed method of extraction involves plunging electric heaters into the ground and bring the rock to a very high temperature and leave it in this condition for three to four years. During this time, the perimeter would be frozen to seal off water and contain the kerogen. To accomplish this task, about 1,200 megawatts of electricity (about the size of one nuclear power plant) would be required to produce 100,000 barrels of kerogen per day. Needless to say, this process is more than twice as expensive to produce oil from oil shale than from oil sands.

Because the U.S. Government owns 80 percent of the oil shale land in the tri-state area of the Green River, previously mentioned, three years ago the U.S. Congress ordered the Bureau of Land Management to accelerate leasing for research and development. Shell Oil was in the running for three of seven 160-acre leases that were considered for approval by the summer of 2006. If follow-up developments succeeded, they would have been converted into a commercial plot of 9 square miles. The bureau also conducted an environmental assessment and weighed how much of a royalty break to give early developers.

Natural Gas and Methane Hydrates

Finally, with regard to carbon based fuels, natural gas and methane—a pure form of natural gas—can be pressurized or liquefied and used as a portable fuel to power vehicles. Typically the natural gas is pressurized to 3,600 pounds per square inch pressure in heavy tanks with about five equivalent gallons of natural gas contained in a 150 pound tank. If 15 equivalent gallons were required, the tanks and mounting apparatus would weigh in at about 500 pounds and take up about half of the trunk space of a typical automobile. A car that weighs 3,500 pounds would then weigh 4,000 pounds.

The U.S. currently consumes about 21 trillion cubic feet of natural gas per year (three percent being imported), and that is expected to increase at an annual rate of about 2 percent per year.

The United States Energy Information Administration (EIA) estimates that the U. S. has about 1,190 trillion cubic feet of proven and unproven reserves. If the 2 percent increase per year prevails, the U. S. will run out of natural gas by the year 2040. If we use only the proven reserves of 282 trillion cubic feet we will run out by the year 2018.

If natural gas were to be used as a transportation fuel, the U. S. would run out of the proven and unproven reserves by about 2031-23 years from now.

To extend the 2031 run out date, T. Boone Pickens has proposed that we use wind power to supplement natural gas usage in stationary power plants. This plan does give us a short term solution to wean the United States off imported oil, and from that standpoint alone the plan has merit. But, like any plan that relies upon non-renewable resources, its life expectancy will be short lived. Besides—as previously discussed—we have the option of making synthetic fuel from coal, and not require an interim conversion of vehicles to natural gas.

In addition the Pickens plan could have unintended consequences, such as the United States converts to natural gas for its transportation fuel, and the rest of the world converts to hydrogen. The U. S. would be behind the learning curve and need to catch up—who knows what this would do to our U. S. automotive industry. A *worst case scenario would be that if the rest of the world also converts to natural gas, for its transportation fuel, the countries that would benefit most would be Russia and the Middle East. Russia and the Middle East currently have about 73 percent of the world's proven natural gas while North America has only 5 percent. The result could be to make the United States, and other countries, eventually dependent on Russia and the Middle East for its transportation fuel!!*

Therefore, it makes sense to continue using natural gas in the way that it is currently being consumed—electric power generation, industrial usage, and home heating. If we are going to retrofit vehicles and develop a new fuel infrastructure let's do it for hydrogen—read subchapters 3.5 and 3.6 to see how this can be done. In the meantime aggressively pursue drilling for domestic oil and producing synthetic fuel from coal.

With regard to methane hydrates, some people have assumed that there are vast quantities of methane hydrates that can be extracted and used in the future. Since the 1970s, methane has been discovered in ice that is lying on, or a few hundred feet below, the deep ocean floor off many coastlines around the world. At first, most experts agreed that the total carbon in methane hydrates was twice that in all the world's known coal, oil, and natural gas combined. Recent estimates, however, are that the total carbon is one-half to one-tenth of all world's known coal, oil, and natural gas. This is still a significant amount that would equal two to ten times that of the world's estimated natural gas reserves.

The extraction of methane hydrates poses some significant obstacles, however, that still need to be addressed. For instance, destabilizing the ocean floor that supports drilling platforms, and sudden releases of methane that is about three thousand times more effective, as a greenhouse gas, than carbon dioxide.

If methane, from methane hydrates, becomes economically available, it would constitute only 15 to 50 percent of the known fossil fuels (coal, oil, etc.). Their use as a transportation fuel would only delay the above-mentioned transportation fuel crisis and exacerbate the situation regarding the current increasing use of natural gas.

Hydrogen

Hydrogen is the most abundant element in the universe and, when made from wind and water, is virtually an unlimited transportation fuel resource. Currently the most common, and economic, way to manufacture hydrogen fuel is from natural gas, and coal, using high temperature steam. Except for the nuclear option, to be discussed later, the by-products of the heating process "potentially" produce the greenhouse gas, carbon dioxide, and other atmospheric polluting gasses. For this reason I am not going to describe, these methods. As illustrated below, my proposed process uses electricity derived from wind and solar energy to create hydrogen gas from water—using electrolysis. The resulting hydrogen gas is then converted into liquefied hydrogen—using a cryogenic freezing process.

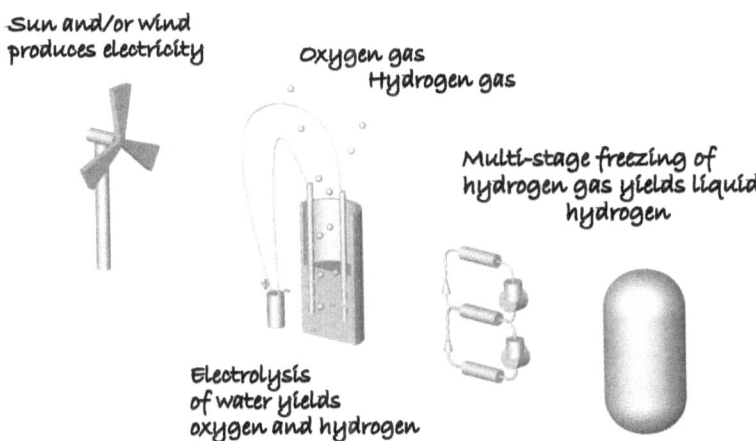

Sun and/or wind produces electricity

Oxygen gas
Hydrogen gas

Multi-stage freezing of hydrogen gas yields liquid hydrogen

Electrolysis of water yields oxygen and hydrogen

Because the proposed process uses sun, wind, and water, its quantity is virtually limitless and cannot be measured in equivalent Saudi oil reserves. However, the manufacture of hydrogen does require large amounts of land area (or ocean area) and production equipment.

It is the main purpose of this book to show how desolate land that is not used for agriculture can be optimized and how the cost of hydrogen fuel-producing equipment can be minimized. As a result, it will be shown that hydrogen fuel production—from wind and water—can be the least expensive mitigation method to limit the impact of oil depletion and the "peak" oil crisis. The proposed hydrogen fuel production process will be shown to be the least damaging to the environment, and when used as a fuel, emits water vapor (i.e., a closed water-to-water cycle).

Hydro and Geothermal

You may be asking, why not use hydroelectric or geothermal to produce hydrogen fuel and/or electric power? The answer is that we are using hydroelectric and geothermal power, but the United States and world energy consumption needs are so great that the availability of hydroelectric and *conventional* geothermal sites fall far short of consumption requirements.

Currently, nonconventional versus conventional geothermal energy production is being given serious consideration. Conventional geothermal uses steam from reservoirs close enough to the earth's surface to be easily tapped to produce electric power. These sites are generally beneath scenic treasures like Yellowstone National Park or on Native American reservations.

Nonconventional geothermal means extracting the heat energy that is found in dry rock about three miles or more underground. To do this, water would need to be pumped into the ground such that it fractures the hot rock and redistributes itself to a number of adjacent extraction wells. If this resource could be tapped, just 2 percent of the available underground heat in the United States could provide nearly two thousand times the power that the nation now consumes each year. As it is with extracting energy from the sun or the wind, the problem is money. Current estimates range from between a competitive 10.5 cents per kilowatt-hour to a sky-high $1.05 per kilowatt-hour.

Summary

While conservation, energy efficiency, ethanol, oil sands, synthetic fuel from coal, natural gas, and oil shale can help reduce the threat of the oncoming depletion and "peaking" of oil, we would be continuing to rely upon polluting hydrocarbon fuels that impact our environment by polluting the atmosphere and *could* lead to global warming. These approaches can be taken—but even they have time limitations that only delay what we can do now—convert to a hydrogen transportation fuel infrastructure.

From what has been presented so far, it doesn't take a rocket scientist to realize that carbon-based "fossil" fuels are not a long-term solution. If all of the above-mentioned oil sand, coal, and oil shale was converted into transportation fuel, and added to the currently known world oil reserves, they would still be depleted within a relatively short period of time. The current 2 percent per year exponential rate of world oil consumption (controlled at 2 percent by using ethanol, conservation, and energy efficiency) would deplete this amount of oil—or fuel—by the year 2067. And this estimate does not take into account the feasibility of producing this amount of fuel from the aforementioned sources or the reduced production rates resulting from the "peaking" oil situation.

And while you may not be concerned about what may happen fifty-nine years from now, it would be irresponsible to our children and grandchildren to leave them in a potentially unstable and polluted world. A world that also had a regressive economy because of our ineptitude, lack of vision, shortsightedness, and greed. *For this generation to ignore using clean burning hydrogen fuel now—especially if its supply is almost inexhaustible and may be the least expensive way to provide fuel—may, at that time, be considered the greatest tragedy in recorded history.*

Additionally, if you think that the above analysis applies only to the rest of the world, and not to the United States, it would be irrational to assume that the United States could continue to fuel itself, while the rest of the world runs out of fuel.

2.3 Can Nuclear Save the Day?

Fission Nuclear

The promise of abundant and cheap nuclear power has yet to be realized. The uranium fuel used to power current fission nuclear electric power plants is not abundant, and because of fuel processing costs, safety measures, and radioactive waste disposal, it is not cheap. It is, however, a way to produce pollution and carbon-free electric power. It can also be used to produce hydrogen fuel and/or charge battery-powered vehicles for transportation.

With conventional fission nuclear power, we need to first remember that this method of power generation suffers from long—and short-lived radioactive waste and the potential for a "meltdown."

There are ways to deal with long-lived radioactive waste like transporting it—in a vitrified "glass" form—into the sun. Short-lived waste can be stored in geologically stable underground sites—like salt mines.

A meltdown is an uncontrolled atomic chain reaction that can occur in a reactor if improper operating procedures are applied. And its consequences can be devastating if it is not properly contained and extinguished. The infamous partial meltdown at the Russian Chernobyl nuclear power plant was the result of improper actions on the part of the plant operators. And because the plant did not have a domed concrete structure to contain the release of radioactive materials, its result was disastrous. In fact it was the Chernobyl incident in Russia, and the "not so lethal" Three Mile Island incident in the United States, that finally turned public opinion against continued use of nuclear power in all of the world's countries—except France. However, current plant design features, computerized operating systems, and redundant safety mechanisms make the meltdown event statistically impossible. And with the inclusion of a containment building with self-extinguishing features, the public should not be fearful.

There is also the potential of making an atomic bomb from very highly enriched fissionable nuclear fuel. Making this fuel is an extremely difficult process that probably could not be done by a terrorist group without being noticed. It would take the dedicated effort of a country—like Iran or North Korea—to make it a possibility. In fact it would be easier to steal a nuclear bomb than make one, and there are no nuclear bombs or bomb-making materials that can be stolen from a nuclear power plant.

To provide the reader with a greater understanding of how a fission nuclear electric power plant works, we first need to understand the nuclear chain reaction process. A fission nuclear reactor produces heat energy by splitting a fissionable uranium atom U-235 using neutron bombardment (see figure 2.3.1).

Courtesy of NEI: The Nuclear Fuel Story

Figure 2.3.1. Nuclear Fission Process

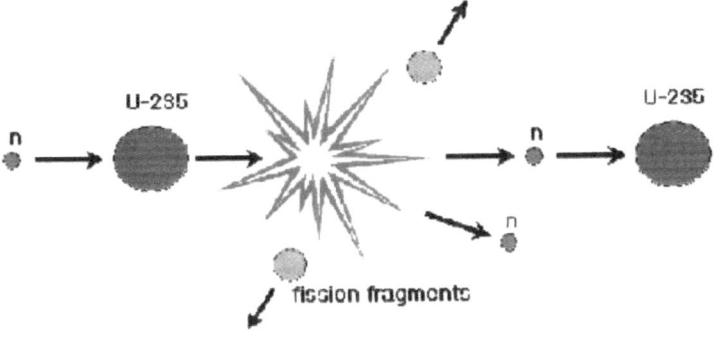

When the U-235 atom is split, it gives off neutrons that split adjacent uranium U-235 atoms to produce a chain reaction that must be controlled in the reactor. Fissionable U-235 uranium is required for this process. And since mined uranium U-238 (yellow cake) contains only a small fraction—less than 1 percent of fissionable U-235—it needs to be enriched to a higher fraction—3 to 5 percent—before it can be used as a nuclear fuel.

During the process of "splitting," there is a resulting small reduction in matter (mass) that is converted into energy, according to Einstein's famous equation $E = MC^2$. In this equation, energy (E) equals the change in mass (M) times the speed of light (C) squared. So even though the change in mass is small, the resulting energy is large. In fact, one pound of 3 to 5 percent fissionable uranium can produce as much electricity as 12,000 pounds of coal or 1,200 gallons of oil.

The heat energy produced by the controlled nuclear chain reaction is used to heat water and become steam. From this point on, the power plant creates electricity like any other electric power plant. The steam expands though a turbine, which turns an electric generator—similar to the alternator that charges the battery in your automobile.

Conventional fission nuclear power currently supplies one-sixth of the world's electricity. In spite of the growing pains of Chernobyl and Three Mile Island, fission nuclear power plants have demonstrated remarkable reliability and safety. Additionally, the problems of nuclear waste disposal is being addressed in countries around the world like Finland, which is currently developing what appears to be a very effective underground nuclear waste facility. Therefore, I support expanding the number of fission nuclear power plants; however, it takes a long time to build nuclear plants, and uranium is not an unlimited resource.

At this point, it should be said that France currently has fifty-nine fission nuclear reactors in operation at twenty sites—including one research breeder reactor.* As of 1996, France had 60,000 megawatts of installed capacity without one major problem. In addition, France is able to build a nuclear plant in about five years. In contrast, because of regulatory control, the United States has taken as much as fifteen years.

Fusion Nuclear

As we look to the future, it is possible that one day we will have low-cost fusion nuclear power. With fusion nuclear power, we may be able to generate abundant and cheap electricity *and* hydrogen from water. The nuclear fusion process involves fusing hydrogen atoms together to produce helium, rather than splitting uranium atoms—as previously described. It is the process that energizes the sun.

Fusion nuclear power plants will be able to produce vast amounts of energy primarily from an element contained in water called deuterium, and another element derived from the world's abundant supply of lithium—called tritium. The process is safe from a meltdown standpoint, and there are no radioactive materials produced.

Figure 2.3.2 is an illustration of the International Thermonuclear Experimental Reactor (ITER). If construction begins as scheduled in 2009,

---

* A breeder reactor can create its own fuel by converting nonfissionable uranium (a byproduct of the uranium enrichment process) into fissionable plutonium. During the process of making electric power, the breeder reactor core will contain nonfissionable uranium U-238. This U-238 material will capture neutrons that convert it into fissionable P-239—plutonium. The plutonium can then be used to power other reactors. In addition, because there would be vast amounts of U-238 tailings from the enrichment process, the breeder reactor could supply electric power, to the world, for hundreds of years, even with a 2 percent exponential increase in the rate of power consumption.

the $12 billion reactor should begin operation in southeastern France in 2016.

Figure 2.3.2. Cutaway View of the ITER Fusion Nuclear Reactor

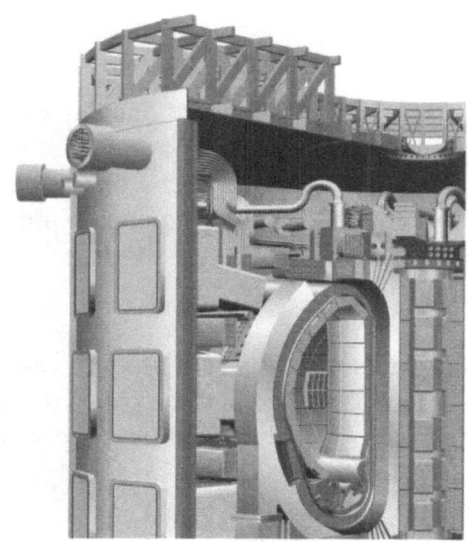

Although it will not add a single watt to the electric grid, the aim of the 500-megawatt ITER reactor is to demonstrate large-scale fusion of the hydrogen isotopes, deuterium, and tritium to generate ten times as much energy as it consumes. A second objective is to "breed" tritium by shooting high-speed neutrons into a surrounding blanket of lithium. A third goal is to integrate the wide range of technologies needed for a commercial fusion plant.

Other fusion reactors in progress are as follows:

China	EAST project	On line 2006
India	SST-1 project	On line 2006
Korea	K-Star project	On line 2008
United States	NIF project	On line 2009
Japan	NCT project	On line (unknown)

Based upon research over the past twenty years, scientists now claim that there is no doubt that fusion will work. The question is its economic practicality and how quickly it can be moved from its experimental form to large-scale commercial reactors. Estimates currently range from twenty-five to fifty years.

With all of the above being said, fission nuclear could be a partial answer to the U.S. electrical energy needs—especially with a breeder reactor that would dramatically reduce (or eliminate) the need for mined uranium.

My Opinion

It is my opinion that it is realistic to assume that fusion reactors will be operating in the not-too-distant future. If so, it would make sense to continue spending research money on fusion reactors and to begin spending research money on fission breeder reactors so that we don't put all of our eggs in one basket.

In the meantime, the United States should restart, and speed up, a fission nuclear program and build as many fission nuclear plants as possible until fusion nuclear and/or breeder fission reactors are available. Keep in mind that without the breeder reactor, the uranium fuel supply will quickly become a limited resource. However, it is unrealistic to believe that enough 1,000-megawatt nuclear power plants can be built in a time frame that will allow them to be used for the generation of hydrogen fuel. On the other hand, if the reactors were modularized—as they are on nuclear-powered ships—we might be able to do it. I will discuss this possibility in more detail in subchapter 4.1.

At this point, you may be asking, why not produce *all* electric power and hydrogen from wind and solar and abandon nuclear power? The answer is that the wind doesn't blow all the time, and the sun doesn't shine all the time. Therefore, it is necessary to be able to produce electric power when it is required. And that is where supplemental, consistently available electric power (from natural gas, coal, and nuclear) is needed to support the hydrogen *transportation* fuel infrastructure.

One may also ask, why not use nuclear power to produce electricity that can be used for battery-powered vehicles rather than use hydrogen? Without going into a long discussion, at this point, about battery-powered vehicles, suffice it to say that there are two broad categories of battery-powered vehicles. The first being an all-battery vehicle that has its batteries recharged overnight—the plug-in; the second is a hybrid vehicle that recharges its batteries while driving—using a small internal combustion engine—this type of hybrid can also have a plug-in option.

For the all-battery vehicle and the hybrid, there are several types of batteries currently being used and new batteries being developed, which I will discuss later. For the all-battery vehicle, it can be said that they suffer from too much weight and space being used to propel them for a short distance relative to current gasoline-powered vehicles. On the other hand, the hybrid vehicle holds a lot of promise and trumps the all-battery vehicle, especially when it is equipped with a hydrogen-powered charging engine. As one person has said, "Battery-powered cars are for carrying batteries, but not very far and not very fast, or else they would have to carry even more batteries." More on battery-powered vehicles will be presented later.

2.4 Antimatter

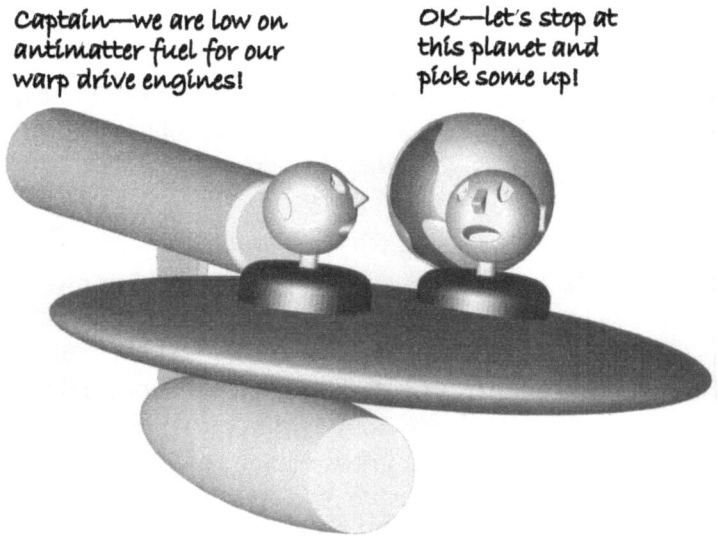

Although antimatter reactors[*] are used on the Starship *Enterprise* (*Star Trek*), and cold fusion (room temperature) was seriously debated in the 1990s, they are probably not something we can count on in the foreseeable future. In the meantime, let's look at more realistic current and future renewable-energy technologies.

Current use of solar, ocean tides, waves, and thermal gradients are projected to be too expensive when compared to other alternatives—but keep them all in mind when used in combination (piggybacked) with wind *power*—more on this later. Wind power is coming into its own as the fastest growing form of electric power generation today and forms the basis for the primary energy source discussed in this book. However, the current form of wind power generation is the propeller-and-pedestal wind turbine, which, in my opinion, is projected to be too expensive and utilizes too much land area—more on this later.

A look into the future reveals an assortment of new ideas that may or may not be realized. Here is a description of a few of these ideas:

High-Altitude Wind

At 15,000 feet, there is a jet stream between 20 and 40 degrees latitude in the Northern Hemisphere. This jet stream is the mother lode in wind power—generating between 5,000 and 10,000 watts per square meter. If this wind power could be captured, it could provide consistent and reliable electric power. At present there is at least one design on the drawing boards at Sky WindPower (see figure 2.4.1). The autogiros shown in this illustration use powered counter-rotating blades to rise above 15,000 feet and then switch to a generating mode. Computers adjust the pitch of four blades to maintain the craft's position and altitude.

---

[*] A kilogram of antimatter and matter would, through their mutual annihilation, produce about half as much energy as all the gasoline burned in the United States last year. Since there are no natural sources of antimatter, it would have to be synthesized. But the most efficient antimatter producer in the world is the particle accelerator at CERN, near Geneva, and it would have to run nonstop for 100 trillion years to make one kilogram of antiprotons.

Figure 2.4.1. Autogiros Designed by Sky WindPower

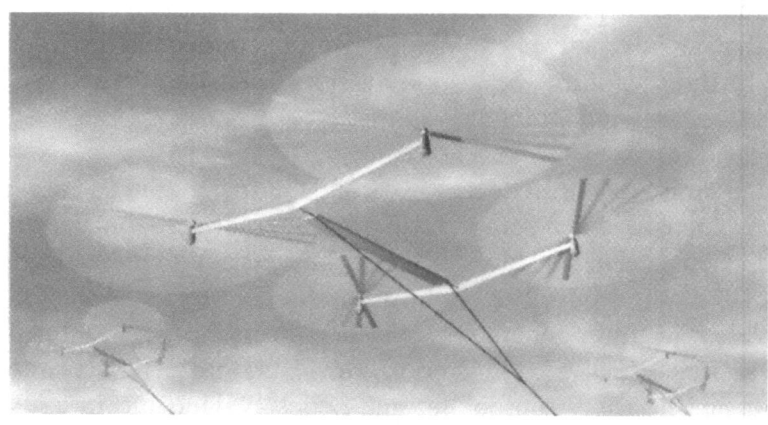

Solar from Space

The average intensity of sunlight in space is eight times as strong as it is on the ground. If this sunlight could be harnessed, it could direct microwaves to a land-based converter that could produce electricity at about 90 percent efficiency. See NASA's proposed version of this technology in figure 2.4.2:

Figure 2.4.2. NASA's Giant Solar Collector in Geosynchronous Orbit

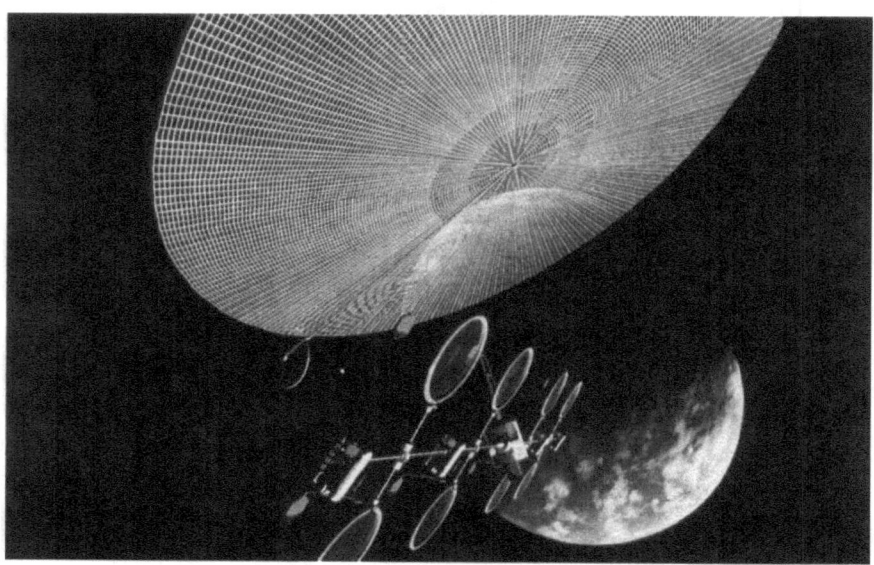

This giant solar collector would work day and night, in any weather, to intercept sunlight and transmit its energy, in the form of microwaves, to a ground-based antenna for delivery as electric energy to the grid.

Tides, Waves, and Ocean Currents

A surging ocean is a virtually untapped energy source. Figure 2.4.3 illustrates a tide farm proposed by Marine Current Turbines. The turbine array would be capable of closer spacing than conventional wind turbines and would be comprised of rotors up to sixty-six feet in diameter. The rotors would sap the energy from tidal currents and have the capability of surfacing for servicing.

Figure 2.4.3. Tide Farm Planned by Marine Current Turbines

Another method of using the ocean's power is the conversion of wave energy into electric power as illustrated in figure 2.4.4. Wave energy devices, made by Ocean Power Delivery, derive electrical power by the flexing motion at their joints as waves pass underneath. Because the machines dive into oncoming waves, they are capable of surviving high seas that accompany intense storms.

Figure 2.4.4. Wave Energy Device by Ocean Power Delivery

Another version, similar to the tidal version, could be used to tap the energy contained in the worldwide ocean currents.

Other Methods

Other methods being evaluated are ocean thermal gradients (to be discussed later), nanotech solar cells (to improve the efficiency of current photovoltaics), a global SuperGrid (to transmit electricity around the world from solar and wind farms using superconducting transmission), and designer microbes (superefficient biological processing to produce hydrogen and/or ethanol).

It is not my intent to evaluate each of the above energy-producing methods, but as you can see, there are many electric energy-producing ideas that may alter or modify the wind and solar* methods proposed in this book. However, the result would be the same—electric power to enable a hydrogen fuel infrastructure from renewable resources.

2.5 Going to Abilene

Many years ago, in a small town in Kansas, the congregation of a small church decided to have a summer picnic. Everyone was to bring a dish of food that would be shared by the others. All was going according to plan. The food was great, and when the meal was finished, the adults and children had a good time playing horseshoes, baseball, and other games. During this time, one of the church leaders suggested they all get in their cars and drive to Abilene—a larger town that was about two hours away. Relative to

---

* At first glance, you may think that the solar/microwave proposal is a great idea, and maybe it is—especially when placed in a geosynchronous earth orbit to provide continuous power to the above-mentioned global SuperGrid. However, the National Aeronautics and Space Administration (NASA) estimated the cost for the first system to be $305 billion (in 2000 dollars). Unless we can learn how to cheaply put large payloads into space, this proposal will remain on the drawing boards. It may be interesting to note that at one time, NASA considered catapulting minerals from the moon (one-sixth of the earth's gravity) to build and operate a space-based solar collector—more on this in subchapter 4.5.

their small town, Abilene was a big town with a lot to see and do. The word spread around the congregation, and everyone agreed they would all go.

It was hot that day, and it had not rained for many weeks. The cars of that time did not have air-conditioning and were prone to having flat tires and other mechanical problems. The road to Abilene was not paved, so with a convoy of many cars, there would be a lot of dust. And since it was so hot, everyone needed to keep their windows open.

As expected, there were car breakdowns during the trip, which meant the convoy needed to stop to lend a hand and to keep everyone together. So rather than taking two hours to reach Abilene, it took about four hours. And since it was midafternoon when they left their picnic grounds, it was very late in the day when they arrived at their destination. Most of the stores were closed, and the restaurants were full. Needless to say, everyone was hungry, but they needed to start back before dark. By this time, the children were not happy, and when the children were unhappy, so were the adults.

The return trip home was just as frustrating—a hot dusty road, more breakdowns, and crying children. During the trip, a high wind started to blow across the cornfields, followed by a heavy downpour of rain, which caused poor visibility, slow driving, and lots of mud puddles. Then one of the cars had a flat tire. Since no one had thought to bring an umbrella, the tire repair was most uncomfortable.

While fixing the flat tire, one person asked, "Why did we go to Abilene when everything was just fine at the picnic area?" Another in the group answered, "We went to Abilene because we thought that everyone else wanted to go." Others then started asking the same question, "How did we end up in Abilene when nobody wanted to go?" And the answers were all the same—"We went because everyone else wanted to go."

At the time of the picnic, nobody thought to ask the question, "Why does everyone want to go to Abilene—especially when it is getting late, and everyone was having a good time?"

What does this story about the trip to Abilene have to do with what we have been discussing so far?

From my perspective, the transportation fuel path that we have been taking can be compared to the road to Abilene. In the 1970s, when the United States had just enough oil to be self-sufficient, we should have seen that we were becoming addicted to foreign oil. At that time, we should have "stopped the convoy" by implementing a hydrogen fuel infrastructure. Instead we kept reaching the point where we now import 70 percent of the oil that we consume.

We are now asking the question, how did we get here—Abilene? Well, we *are* here, but we need to act decisively so that the problem doesn't get any worse. The current fixes are not going to solve the problem, and the sooner we recognize that fact, the better will be our chances of survival, and *survival* is the appropriate word.

Because we are talking about survival, the facts to be presented in this subchapter are so important that many of you may want to put them to memory. Then, when put to memory, they should be acted upon in terms of informing family, friends, neighbors, the media, and government officials.

The title of this chapter should probably be modified to "Hydrogen Fuel Is the Only Answer." When the term *only* is used, it is like never saying never. However, here is the reasoning behind this claim.

To begin, it is important to free yourself from all preconceived notions about what you have read (except for the earlier subchapters of this book) or what you have heard people—including scientists—say. *The fact of the matter is, all nonrenewable carbon-based fuel will be depleted in the not-too-distant future, and the only thing left will be renewable energy, nuclear power, and hydrogen. It is as simple as that.*

Since renewable energy includes ethanol or biofuels, we need to consider how population growth will impact the conflicting need for food versus the need for fuel. Therefore, at some point in the future, a balance of food and energy would create a limit on this renewable energy source. Unless, of course, we are able to convert our desolate nonfarmland into ethanol-producing land (more on this later) or produce cellulostic ethanol from crop residue.

Other forms of renewable energy sources, such as solar and wind, are generally used to create electrical energy. This electrical energy can be stored in the form of hydrogen fuel or to charge batteries. Other storage methods are possible, but are generally too bulky to use as a portable energy source for transportation. Otherwise, electricity needs to be used as it is created.

Realizing that fissionable nuclear fuel is also a limited resource, it must be concluded that the future of nuclear power must be either a fission-type breeder or fusion nuclear.

Unless a breakthrough in battery technology occurs, hydrogen will be the preferred transportation fuel, but perhaps in combination with batteries—a hybrid.

We need a short and intermediate term transportation fuel solution that will quickly free the United States from its addiction to imported oil. In my opinion the intermediate term—3 to 7 years—solution should be to aggressively pursue the coal to synthetic fuel option and drilling for more domestic oil. As a shorter term solution we should reduce our oil consumption through conservation, fuel efficiency, and ethanol/biofuels. Another short to intermediate term solution is to use hydrogen fuel made from coal and natural gas (read subchapters 3.5 and 3.6). In my opinion, an aggressive intermediate term solution will immediately stabilize oil and gasoline prices. I believe this because it will cause oil speculators to assume that future oil prices will go down rather than up.

If we can agree at this point, the only question that remains is, when will we run out of carbon-based fuels, and when will we maximize the use of ethanol (if it continues to be used when hydrogen is available)?

To answer these questions, we first need to reestablish the fact that the world's population is rapidly increasing. As previously mentioned, in 1950 there were 3 billion people; today there are over 6.6 billion people, and the projected number for 2050 is 9.2 billion people (the predominant increase occurring in nonindustrialized countries). Hopefully, this will stabilize with increased worldwide industrialization; however, with more industrialization comes more fuel consumption.

Currently, the world's consumption of oil is increasing at a rate of more than 2 percent per year (recently 2.6 percent). At this rate, the world's oil consumption rate will increase from today's 85 million barrels of oil per day to 197 million barrels of oil per day in the year 2050, and 520 million barrels of oil per day by 2100. And it is my opinion that the 2 percent rate will not be curbed within the next fifty years by conservation, energy efficiency, and the current rate of using alternate or renewable energy sources.* At best, these methods will hold growth at 2 percent rather than the current trend to a higher percentage rate—more on this later.

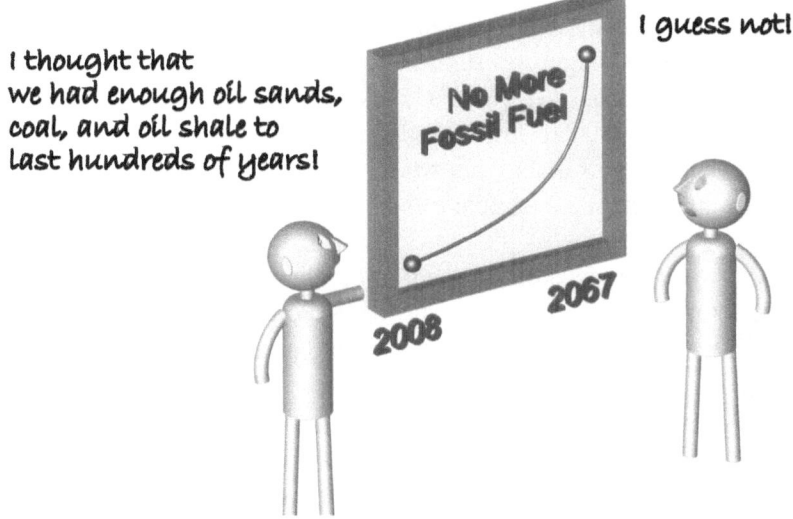

Using the information supplied in the previous subchapters, let's do some simple arithmetic:

---

* To elaborate further, if we reach a point where oil "peaks" and the other fossil fuel supply sources cannot produce fast enough to keep up with demand (which is very likely), then supply and demand pricing will take over and control the rate of consumption. Obviously, this will result in skyrocketing prices that will have a devastating effect on the industrialized world economy. Other actions, like fuel taxes and rationing, will have a similar detrimental result.

The world's oil reserves in the year 2001 was about 1,032,132 million barrels. At the 2 percent annual rate of consumption increase, the world will be out of oil by the year 2028.

The second easiest to recover oil is found in oil sands located mostly in Canada and Venezuela. The world's recoverable-oil reserves are estimated to contain about 485,000 million barrels of oil. At a continuing 2 percent annual rate of consumption increase, the world would run out of oil-sand oil by the year 2036.

If the world's recoverable coal were used to produce electricity *and* synthetic fuel at a 2 percent annual rate of consumption increase of electricity *and* transportation fuel, the world would run out of coal by the year 2061. Note: This fact may seem hard to believe when we have 2,002 trillion pounds of worldwide recoverable coal currently available, but when you calculate current energy consumption, at a 2 percent growth rate, you will see why!

The world's recoverable supply of oil (kerogen) from oil shale is estimated to be 620,000 million barrels. At a continuing 2 percent annual rate of consumption increase, the world would run out of oil shale by the year 2067.

Now assume that conservation, energy efficiency, and ethanol/bio-fuels (which will probably reach a balance between world food and fuel requirements) will be able to maintain the exponential rate of transportation fuel usage at 2 percent rather than go to 3 percent or more.

Since the use of natural gas is also increasing at a 2 percent annual rate for electric power generation, industrial usage, heating, and possibly providing the heat necessary for processing the oil sands, oil shale, and ethanol, it will be depleted in a similar timeframe. However, because of the uncertainty of how much unproven natural gas or methane hydrates will become available its run out date is, at best, uncertain.

Since methane hydrates and natural gas will be needed during this time frame for electric power generation, industrial usage, heating, and possibly providing the heat necessary for processing the oil sands, oil shale, and ethanol, it will also be depleted in the above time frame.

Note: If the entire expected 2050 world population were to consume oil at the rate currently consumed by the United States, the rate would be 637 million barrels of oil per day (compared to the world's current consumption of 85 million barrels per day). *This is more than the projected consumption rate for the year 2100—when calculating on the basis of 2 percent annual growth—and is enough to deplete the current Saudi oil reserves in about one year.*

With all of the above being said, including the preceding subchapters, it should be clear that the United States should begin developing a hydrogen *transportation* fuel infrastructure to replace gasoline, diesel, and aviation fuels. Conservation, energy efficiency, oil sands, oil shale, synthetic fuel from coal, natural gas, ethanol, and possibly methane hydrates are, at best, temporary solutions to the problems at hand (and should only be used in the short term while a hydrogen fuel infrastructure is being developed). They should definitely not be thought of as long-term solutions, and the sooner we get on with converting to hydrogen fuel, the sooner we, and the rest of the world, can prosper from the numerous derived benefits that are described later in this book.

Converting to a hydrogen *transportation* fuel economy is a significant part of what this book is all about, and while you may not agree with some of the aspects of my proposals, it is at least a starting point from which further discussion and ideas can take place. *There are shortcomings to using hydrogen as a transportation fuel, but rather than trying to find reasons for not using hydrogen, it is time to begin finding economical and safe ways to use it.*

As previously mentioned, if we are successful in doing what is proposed in this book, the world will follow. If we don't do it (or something similar), another country will. A good guess would be China or Japan.

My estimated timeline for a 100 percent conversion to hydrogen fuel for transportation is sixteen years. The time period for the elimination of Middle East oil imports would be less than nine years. The nine year timeframe could be significantly reduced with an aggressive short-term program of conservation, energy efficiency, drilling for more domestic oil, use of cellulostic ethanol, and producing synthetic fuel from coal. As a parallel effort I propose converting existing diesel-powered equipment to a hydrogen fuel blend by using hydrogen derived from coal—details of this proposal will be provided later in subchapter 3.6. In addition, I propose that we retrofit a percentage of existing automobiles with dual fuel tanks—hydrogen and gasoline; details will be provided in subchapter 3.5.

Chapter 3

Wind and Water

Where there is a will, there is a way.

—Anonymous

3.1 A Compelling Scenario

Remember the quote, by Lord Kelvin in the last chapter, about applying numbers to show your knowledge about a subject? Well, it is not just numbers, but numbers plus compelling reasoning and logic that will help to convince the skeptic. In this chapter, I will put the hydrogen argument into numbers that may or may not convince the skeptic, but it is a starting point for rational discussion. Furthermore, for simplicity, I am going to use *feet* instead of *meters*, *miles* instead of *kilometers*, *pounds* instead of *kilograms* (except when referring to a kilogram of hydrogen), and *Fahrenheit* instead of *Celsius*.

Imagine, if you will, that you are at a neighborhood block party, and someone says to you, "The price of gasoline has just gone up by ten cents per gallon—when are these price increases going to end?" You respond by saying, "Well, I just read a book that said we could have hydrogen fuel for about the same price as gasoline if the United States would only begin the process of developing a hydrogen fuel infrastructure." Surprised by your response, this person then says, "Hydrogen is too dangerous to be used as a fuel, just look at what happened to the *Hindenburg*." You then explain how hydrogen is safer to use than gasoline in the event of accidents and spills and that the *Hindenburg* accident actually helped to prove this point. Amazed by your response, this person then challenges your assertion. After you explain the *Hindenburg*'s numbers and facts, this person then says, "Well, if hydrogen fuel is so safe, and can be made for about the same price as gasoline, why haven't we done it before?" At this point of the discussion, you need to have a clear understanding of how hydrogen fuel can be economically produced from wind and water and then delivered to a vehicle's fuel tank. It is the purpose of this subchapter to give you the basic knowledge of how to answer this question. And depending on how knowledgeable your conversationalist is, the questions can get more and more complex. In this case, you will need to read and understand the rest of this chapter and the appendix to be on firm ground. But don't worry, the information that I will present will be made entertaining and easy to understand. You will be surprised at how hydrogen fuel savvy you will become.

As the old saying goes, "A picture is worth a thousand words." Because this is a very true saying, I am going to do some simple arithmetic using a simplified illustration of a process that shows how hydrogen fuel can be economically produced and delivered to your vehicle fuel tank.

The following was tested on an eleven-year-old boy, who was in the fifth grade, to see if it was too complicated to understand. As I showed and explained the sequence of processes to this boy, it was clear that he not only understood the concept, but he could explain it in his own words after one time through.

However, before we get to the process illustration and accompanying arithmetic, it is important for you to know and understand the following six points:

A kilogram (2.2 pounds) of liquefied hydrogen is equal to about one gallon of gasoline in energy content.*

A kilowatt-hour is a measure of electric power, just as you will see on your home electric bill. It is the amount of electricity—in kilowatts delivered—multiplied by the number of hours that it has been used.

Water is made up of hydrogen plus oxygen—you know H_2O—and both hydrogen gas and oxygen gas can be separated from the water molecule using electric power—kilowatt-hours. To make the hydrogen more like gasoline, we are going to freeze it so that it becomes a liquid. This process takes more electric power—kilowatt-hours.

Hydrogen gas can be transported using a pipeline, but it takes compressors to propel it along the pipeline. As it is propelled along the pipeline, friction, caused mainly by the pipe's inner surface, causes heat energy to develop that results in a loss of energy. This energy loss can be expressed as kilowatt-hours.

Electricity can be transported using wires; however, one form of electric transmission is superconductivity—which we will discuss later. The electricity being transmitted along any form of transmission line creates heat energy. This then results in a loss of electric power, which can also be expressed as kilowatt-hours.

---

* Energy content is the amount of stored energy contained in a specified quantity of fuel. This energy can be expressed in many ways, such as British Thermal Units (BTUs) or kilowatt-hours. For instance, one kilogram of hydrogen contains about 117,000 BTUs of stored energy, while gasoline contains 39,400 BTUs per kilogram. At 3.04 kilograms per gallon of gasoline, a gallon of gasoline contains 119,800 BTUs.

Now let's examine the overall process for producing and dispensing hydrogen fuel as described in the illustration below:

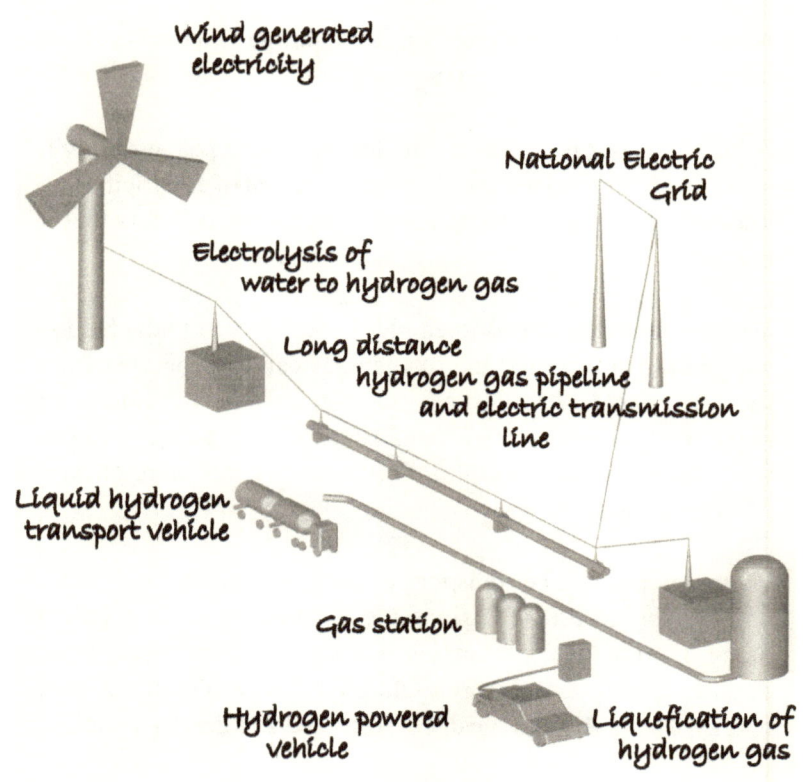

As you look at this illustration, notice in the upper left-hand corner a conventional wind-activated electric power generator. At this point, just consider that one wind generator is capable of gathering 5,000 kilowatts (or 5 megawatts) of electrical energy from the wind—more on this later.

The next thing that you see, as you go from left to right, is an electrolyser. This device converts water into hydrogen and oxygen using electric power (kilowatt-hours)—we will discuss more about this later. However, for the electrolyser shown, it takes 55 kilowatt-hours of electric power to make 1 kilogram of hydrogen gas, and we will be using 10,000-kilogram-per-hour electrolysis machines to do this.

OK, now notice a combined long-distance pressurized hydrogen gas pipeline and electric transmission line. It "branches" off to a number of central processing stations (branching not shown) where the electric power is used to freeze the hydrogen gas into a liquid form. Insulated tanks are then used to store this liquefied hydrogen until it is dispensed to hydrogen delivery vehicles.

The long-distance transmission of gas and electricity results in a loss of energy—kilowatt-hours—and we will assume an 8 percent loss of energy in this process. In addition, fifteen kilowatt-hours are required to make one kilogram of liquefied hydrogen fuel, and we will be using 10,000-kilogram-per-hour processing machines to do this. Notice that a "branch" connection is also directed to the National Electric Grid System—more on this later.

Next, notice that a double tractor-trailer is transporting liquid hydrogen to a gas station. While at the gas station, the transported hydrogen tanks are exchanged for empty tanks. The double tractor-trailer then returns to the central station for refilling the empty tanks. The liquefied hydrogen at the gas station is then dispensed into a hydrogen-powered vehicle fuel tank.

In our compelling scenario, we are going to produce 10 percent of the current United States transportation fuel consumption that is 1.97 million kilograms of hydrogen per hour.* Also, we are going to assume that an average of 140,000 megawatts of electricity is required by the wind-powered generators to produce and transport this amount of hydrogen fuel to its destination.

As shown, this process continues at many central processing stations (about 190 stations), and each processing station distributes hydrogen fuel to many gas stations (about fifty for each processing station) that are located a short distance away (within a 100-mile radius). In special cases, the gas station converts some of the liquid hydrogen back into a gas form for dispensing to hydrogen gas-powered vehicles.

---

*   The United States currently consumes about 21 million barrels of crude oil per day (882 million gallons). Crude oil is refined to make gasoline, diesel, aviation fuel, fuel oil, and other residual oils and gases. The portion that is made into gasoline, diesel, and aviation fuel equals about 472 million gallons per day or 19.7 million gallons per hour. Ten percent of 19.7 equals 1.97 million gallons per hour.

The first thing we are going to do is determine the cost of producing electric power from the wind—remember, we need to produce an average of 140,000 megawatts of electric power, but since our 5-megawatt generators are only producing at 50 percent of their capability (more on this later since we will need a different type of wind-activated electric generator to achieve 50 percent), we need to install 280,000 megawatts of generating capacity. Don't be concerned at this time about the cost numbers being used, since this will also be explained later, but remember that our goal is to produce 1.97 million kilograms of hydrogen fuel per hour. So here we go:

> Oh, one other thing, we are going to add $10 billion for supplemental solar power. This will generate additional electricity and help to drive down solar photovoltaic costs.

Are we OK so far? If not, review the above information again, and it should become clearer as we progress.

Step 1—Cost of Electricity

I will begin by making the assumption that we can generate electricity from the wind for $2,000 per *actual* kilowatt generated. This will be explained in more detail later.

Our next assumption is that we can generate an average of 36,000 *actual* kilowatts per square mile of land or ocean area. Again, this will be explained in more detail later.

With the above assumptions, a total of 3,010 square miles are required to produce 1,970,000 kilograms of liquefied hydrogen fuel per hour and costs $280 billion (3,890 square miles x 36,000 kilowatts per square mile x $2,000 per kilowatt = $280,000,000,000).

At 6 percent interest over ten years, $140 billion = $37.28 billion per year.

Step 2—Cost of Electrolysis

Water is transformed into pressurized hydrogen gas by a process called electrolysis.

A total of 55 kilowatt-hours of electric energy are required to produce one kilogram of hydrogen (which equals one gallon of gasoline).

A total of 10,000 kilograms (of hydrogen gas) per day machines are used (417 kilograms per hour).

A total of 1.97 million kilograms of hydrogen gas is required per hour, but the maximum capability of the electrolysis machines needs to be twice this amount—or 3.94 million kilograms per hour (because the wind generators are capable of operating at 100 percent efficiency—rather than the assumed 50 percent efficiency).

Therefore, 9,448 electrolysis machines are required—let's use 9,500 machines (3,940,000 kilograms per hour/417 kilograms per hour per machine = 9,448 machines).

Step 3—Cost of Hydrogen Gas and Electricity Transmission

The total energy lost during long-distance transmission of hydrogen gas and electricity (for liquefying the hydrogen gas at the central processing station) = 8 percent. Therefore, for the maximum 280,000 megawatts of electric power produced (3,890 square miles x 36 megawatts x 2 for 100 percent operating efficiency = 280,000 megawatts) at the wind power-generating site, only 258,000 megawatts are available to produce hydrogen fuel (280,000 megawatts x 0.92 percent = 257,858 megawatts).[*]

The cost of the pipelines and electric transmission lines = $40 billion.

At 6 percent interest over ten years, the cost of $40 billion = $5.33 billion per year.

Step 4—Cost of Hydrogen Liquefaction

Pressurized hydrogen gas is transformed into liquefied hydrogen through a multistage refrigeration process.

---

[*]   For simplicity, we are going to take the transmission energy loss as if the hydrogen electrolysis and liquefaction processes were both located at the central processing station.

Besides the 55 kilowatt-hours needed for electrolysis to produce the hydrogen gas, an additional 15 kilowatt-hours of electric energy is required to liquefy one kilogram of hydrogen—therefore, 70 kilowatt-hours are required to produce one kilogram of liquefied hydrogen from water (55 kilowatt-hours + 15 kilowatt-hours = 70 kilowatt-hours).

A total of 10,000 kilograms per day (417 kilograms per hour) liquefaction equipment is used to liquefy hydrogen gas.

Since 9,500 electrolyser machines are required, an equal number of equal capacity liquefaction machines are also required.

At $10 million per electrolyser and $2 million per liquefaction machine, the cost of 9,500 machines is $114 billion.

At 6 percent interest over ten years, the cost of $114 billion = $15.19 billion per year.

Step 5—Cost of Hydrogen Storage and Transmission

Tractor-trailers deliver modular tanks filled with 2,500 kilograms of liquefied hydrogen fuel to gas stations located within a radius of 100 miles from the central processing and dispensing station.

Robotic equipment loads and unloads modular fuel tanks, and empty tanks are returned to the central station for refilling.

The cost of the modular tanks, tractor-trailers, robotic equipment, dispensing equipment, central station storage, and "buffer" tanks is $1 million per gas station. Total cost of gas station operations is $9.5 billion (we will assume 190 central stations x 50 gas stations per central station = 9,500 gas stations x $1,000,000 per gas station = $9,500,000,000).

At 6 percent interest over ten years, the cost of $9.5 billion = $1.27 billion per year.

Step 6—Total Cost of Producing and Dispensing Hydrogen Fuel

The yearly cost of operations, maintenance, water, land leases, and the addition of supplemental solar power is estimated to be $2.34 billion.

Yearly operations and solar cost	= $2.34 billion
Yearly wind-energy machines cost	= $37.28 billion
Yearly electric/pipe transmission cost	= $5.33 billion
Yearly electrolysis and liquefaction cost	= $15.19 billion
Yearly gas station equipment cost	= $1.27 billion
Total cost per year	= $61.4 billion

Note: If 1,970,000 kilograms of liquefied hydrogen fuel per hour x 8,760 hours per year = 17,257,200,000 kilograms of liquefied hydrogen fuel per year, then the cost of liquefied hydrogen fuel = *$3.56 per kilogram* ($61,400,000,000 per year/17,257,200,000 kilograms per year = $3.56).

Step 7—Savings Resulting from Using Hydrogen Fuel

Assuming a future price of $4.00 per gallon of gasoline, the cost to fill 20 gallons would be $80.00—not an unreasonable assumption when you consider the information presented in the previous chapter.

For the same performance hydrogen-fueled "internal combustion engine" vehicle, we will get 25 percent more miles per equivalent gallon of hydrogen fuel—therefore, at $4.63[*] per kilogram of hydrogen fuel, the cost to fill 16 equivalent gasoline gallons = $74.08.

---

[*] The $4.63 per kilogram = $3.56 x 30 percent markup for sales costs, profits, and federal/state taxes (Note: In many cases, this markup is 40 percent, but if our government wants the hydrogen fuel economy bad enough, it could reduce its tax rate—*I have assumed that this is the case*).

For the same performance hydrogen fuel cell vehicle, we will get 100 percent more miles per equivalent gallon of hydrogen fuel—therefore, at $4.63 per kilogram of hydrogen fuel, the cost to fill 10 equivalent gasoline gallons = $46.30.

The hydrogen-fueled vehicle savings = $5.92.
The hydrogen fuel cell vehicle savings = $33.70.

So there you have it. The arithmetic shows that the price of hydrogen fuel is mainly a result of the cost of equipment used to produce it—primarily the wind-powered electric generators. As you know, the wind energy is free, and the water, operations maintenance, solar power supplement, and land lease costs are relatively small. However, you can be sure that if the assumed costs are what current manufacturers say they are—especially the wind-powered electric generators and electrolysis machines—then the cost per hydrogen equivalent of one gallon of gasoline would not look as good as I have shown. *And that is the problem that misleads many analysts to believe that hydrogen fuel, made from wind and water, costs too much. As a result, they dismiss "water-produced" hydrogen as a viable near-term fuel.*

3.2 Facts can be Confusing

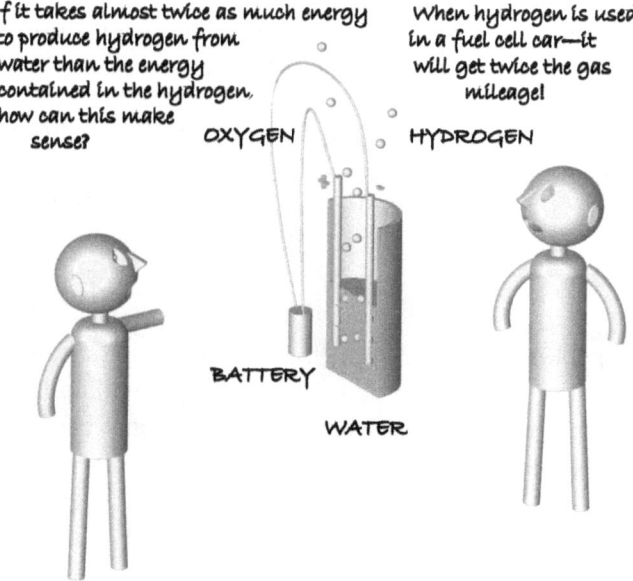

If it takes almost twice as much energy to produce hydrogen from water than the energy contained in the hydrogen, how can this make sense?

When hydrogen is used in a fuel cell car—it will get twice the gas mileage!

OXYGEN

HYDROGEN

BATTERY

WATER

Sometimes "*facts can be confusing*" especially if they are counterintuitive or interfere with what you believe. If, at one time, someone told you that we have a three-hundred-year supply of coal, you probably believed this to be true and never gave it much more thought. However, in this book, you have been told that at future consumption rates, coal will last less than twenty-five years. Now that's confusing. How could we be off by so much? There is an old saying, "A little bit of knowledge can be dangerous."

Then there are facts that can be misleading and cause people to make erroneous conclusions. Here is where wisdom and common sense come into play. T. S. Elliot once wrote, "Where is the wisdom we have lost in knowledge?"

Unfortunately much of what has been presented so far, in this book, runs counter to what many people believe or what they have been told. However, as Ronald Reagan once stated, "Facts have a funny way of sticking." Hopefully you will not just accept everything written in this book. There is nothing preventing you from doing your own research—especially with the Internet being available to everyone. But be careful when the theme of what is being presented has political or monitory overtones—remember, follow the money. This includes statements made by legitimate scientists who support their beliefs with scientific facts. Make sure that you research the scientific facts that may support another point of view or theory. And that includes the facts and information presented in this book. By knowing both sides of the story, you will at least have the basis for applying wisdom and common sense.

While reading the last subchapter, you may have developed some questions. If so, I will attempt to preempt your questions by asking some relevant questions and then providing answers. If you are still not satisfied, more answers are provided in the rest of the book, and of course, you can do your own research. Hopefully the questions and comments listed below will answer most of your questions:

Question 1

The analysis shown in the previous subchapter assumes that electric power derived from the wind costs $2,000 per *actual* kilowatt produced. Current propeller-and-pedestal wind turbines produce electric power at more than $4,000 per *actual* kilowatt produced. How can this assumption be made?

Comments

Let's get this issue out of the way first, since more than 60 percent of the final cost of hydrogen fuel is based upon the cost of wind-generated electric power. Assuming the current $4,000 per actual kilowatt, and a 30 percent markup, the price per kilogram of liquefied hydrogen fuel would be $7.43 (as previously stated, one kilogram of hydrogen is equal in energy content to one gallon of gasoline). However, if my proposed advanced mass production, as described in the appendix of this book, were used to make large quantities of 5-megawatt propeller-and-pedestal wind turbines, I would expect that their installed price to be about $3,000 per actual kilowatt. This would result in a fuel price of $5.57. This is still high, but we will double our mileage with a hydrogen fuel cell vehicle. Therefore, we can assume that the price of fuel—based upon hydrogen fuel cell mileage—would be half of $5.57 or $2.79—including taxes and profit.

A potentially better form of wind power technology is the Magenn floating air rotor as illustrated on the cover of this book. The floating air rotor is a lighter-than-air device that rotates with the wind to generate electric power. It is held in place with a tether cable that permits the air rotor to align itself with the wind and transmit electric power to ground level. Helium is currently proposed for the lighter-than-air gas, but there is no reason why less expensive and more available hydrogen gas could not be used (a double-wall air rotor could use hydrogen as a center core gas with helium surrounding it). With this technology, we can expect a mass-produced price of about $2,000 per actual kilowatt of electric power produced. This is because the air rotor operates at high altitudes (a doubling of altitude results in about 34 percent more wind power); therefore, we can assume a 50 percent rather than 25 percent efficiency. Although Magenn's current nonmass-produced price per rated power output is currently higher than conventional propeller-and-pedestal wind turbines, the 50 percent operating efficiency plus mass production make the $2,000 per kilowatt feasible because of its lighter and less expensive materials and less costly deployment costs. As a further benefit, we will use about half as much land or ocean area because of an ability to deploy the lighter-than-air

air rotors in a controlled three-dimensional spacing. Therefore, with a 30 percent markup, we can produce a kilogram of liquefied hydrogen for about $4.63—as shown in the previous subchapter.

Question 2

Hydrogen fuel costs too much—especially when it is made from wind and water.

Comments

Read the appendix of this book to see how these costs can be achieved through advanced mass production.

To put things into perspective, look at the value provided by an automobile, and compare that value to the hydrogen-producing equipment planned for this economic analysis—like an electrolysis machine. If a new—conventionally mass-produced—automobile is priced at $25,000, what do you think a mass-produced, much less complex (10,000 kilograms of hydrogen per day) electrolysis machine should cost if it were mass-produced? For this analysis, we have assumed a value, for the electrolysis equipment of $10 million (at current nonmass-produced pricing, this could be as much as $30 million), which is equal to buying four hundred *$25,000 automobiles. So the question is not what do things cost; instead, the question should be, what should they cost if mass-produced?*

Question 3

How can we economically make hydrogen fuel when it takes more than twice as much electrical energy than the energy contained in the stored hydrogen?

Comment

As shown in the answer to question 1, if we were to use conventional propeller-and-pedestal wind turbines, and hydrogen fuel cell vehicles, the arithmetic shows that we are cost competitive with gasoline even when a greater amount of electricity is used than is contained in the hydrogen fuel—"facts can be confusing."

Question 4

Since it appears that we are wasting electrical energy, why not use this energy to charge highly efficient lithium-ion batteries in automobiles?

Comments

Let's first compare hydrogen to gasoline. One kilogram of hydrogen fuel is about equal in energy content to about one gallon of gasoline—36 kilowatt-hours of energy. However, the hydrogen fuel is twice as efficient in a fuel cell vehicle than gasoline fuel used for an internal combustion engine vehicle. Therefore, one could argue that even though it takes 78 kilowatt-hours of electric energy to make 36 kilowatt-hours of liquefied hydrogen fuel energy, the 36 kilowatt-hours of hydrogen energy—when used in a fuel cell vehicle—is worth 72 kilowatt-hours (36 kilowatt-hours x 2) of equivalent gasoline energy. Thus, making the lost electrical energy issue less of a concern. Also, when we are able to use gaseous, rather than liquid, hydrogen, the energy requirement to transform water into hydrogen is 55 kilowatt-hour plus transmission line losses of about 8 kilowatt-hours. This advantage becomes very important, in the future, when we electrically transform water into hydrogen gas at each gas station.

If the wind-generated electricity were used to charge your battery-powered car, the result would be a more efficient use of the electric energy—if the wind energy were available on demand. However, batteries consume much more space and weight than liquefied hydrogen in a fuel cell-powered vehicle, which results in less "miles per gallon." Thus negating the presumed savings in energy resulting from the direct charging of batteries—"facts can be confusing." I will explain more about this later.

Again—the fuel source of the generated electricity, in this analysis, is the wind (and some solar energy)—it is free and creates no atmospheric pollution. It is only the land, or ocean, area being used to generate electricity, the environmental impacts, and the cost and availability of materials that go into producing hydrogen fuel that should be of any concern. One could argue that all of the wind and solar energy that we don't capture, and use, is wasted.

Question 5

Current 5-megawatt propeller-and-pedestal-type wind turbines are about 25 percent efficient—relative to rated capacity—and, when placed with proper spacing, would result in an average of about eighteen thousand *actual* kilowatts per square mile. Although the Magenn air rotor will cut this to about half, do we have enough land area to produce all of the electricity needed to fuel the United States?

Comments

At 18,000 actual kilowatts per square mile, for 5-megawatt propeller-and-pedestal wind turbines, the required land area to supply the United States with hydrogen fuel is about 77,000 square miles (400 miles wide by 195 miles deep equals 78,000 square miles). If you look at the wind map shown in figure 3.2.2, you will see that most of North and South Dakota have average annual winds in excess of 400 watts per square meter, and 78,000 square miles would occupy about 54 percent of the combined states' land area. However, when spread out to other available sites throughout the country, it is reasonable to use conventional wind turbines.

Courtesy of United States Department of Energy

Figure 3.2.2. Wind Map of the United States

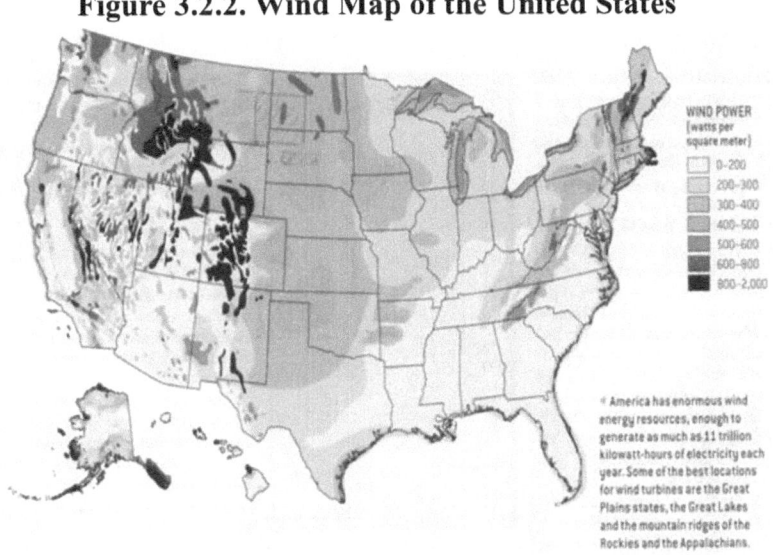

WIND POWER
(watts per square meter)

0-200
200-300
300-400
400-500
500-600
600-800
800-2,000

◄ America has enormous wind energy resources, enough to generate as much as 11 trillion kilowatt-hours of electricity each year. Some of the best locations for wind turbines are the Great Plains states, the Great Lakes and the mountain ridges of the Rockies and the Appalachians.

Using Magenn air rotors for wind power generation, the land area to supply the United States with hydrogen fuel would be about 38,000 square miles (300 miles wide by 130 miles deep equals 39,000 square miles). This is equivalent to about 27 percent of North and South Dakota's land area.

In contrast, the "breakthrough" wind power technology, to be described in subchapter 4.1, will occupy about 11,000 square miles (150 miles wide by 75 miles deep equals 11.25 thousand square miles, or 8.5 percent of North and South Dakota).

Question 6

Some critics may say that superconductivity is an impractical way of transmitting electric power, so why not use conventional high-voltage direct current (HVDC), which is commonly used to transmit bulk electric power?

Comments

OK, let's use HVDC. However, because we already have low-temperature liquid hydrogen available in our process, why not give it some consideration. Besides, who wants more unsightly overhead transmission lines?

To elaborate further, superconducting wires are capable of transmitting five times as much electricity per wire, which means that a 3,000-megawatt HVDC transmission line can be replaced with a much smaller superconducting line that can carry as much as 15,000 megawatts. To emphasize this point, one study showed that 250 pounds of superconducting cable can replace 18,000 pounds of copper wire power cables—a 7,000 percent decrease. If aluminum wire cable were used for the HVDC line, the improvement in weight and cost savings would not be as great, but still significant.

Read the appendix to see a proposed way of integrating the hydrogen gas pipeline and superconducting electric transmission line.

Question 7

The argument is sometimes made that it takes more energy to produce the power-generating equipment than the energy that the power-generating equipment produces. Is this true of the "compelling scenario" proposal?

Comment

Regarding the subject of energy consumed, there is always the argument that more energy is used to create the energy-producing equipment than the energy produced by the equipment (i.e., energy in versus energy out). This argument certainly applied to nonphotovoltaic solar panels that use highly energy-intensive materials like aluminum and glass. And as previously discussed, in subchapter 2.2, in the case of making corn ethanol, a large amount of energy is used in farming and distillation relative to the amount of energy contained in the ethanol fuel. This argument was also used for "fission" nuclear power plants. It was once calculated that the first seven years of the nuclear plant's forty-year life expectancy was spent in recovering its construction energy. However, in our "compelling scenario," the amount of energy used to make and operate the hydrogen generating equipment has been calculated to be less than 2 percent of the hydrogen energy produced.

Question 8

Some critics have claimed that breakthrough technologies are required before proceeding with a hydrogen fuel infrastructure— especially with regard to hydrogen fuel tanks in cars. Is this true?

Comments

No breakthrough technology is required to realize the claims made in the "compelling scenario." However, the wind powered electric generator proposed in subchapter 4.1 would be a breakthrough and needs to be developed. If for some reason this new technology cannot be employed, a fallback position would be to use the Magenn floating air rotor or conventional 5-megawatt propeller-and-pedestal wind turbines. Another fallback position might be to use other forms of renewable electric power, if their economics becomes favorable.

A thorough discussion of hydrogen processing and delivery to your vehicle's fuel tank is provided in the appendix. However, a discussion of fuel tanks for hydrogen-powered vehicles is provided later in subchapter 3.5. As you will find out, no breakthrough technology is required.

Question 9

A number of technical reports have been written that show hydrogen fuel is either too expensive, too dangerous, or requires a breakthrough technology before it is implemented. Therefore, hydrogen is "decades away." Is hydrogen fuel "decades away?"

Comments

Hydrogen fuel is not "decades away" as described in this book. However, a Manhattan Project-type is needed to make it happen in about sixteen years. Whether or not you believe the hydrogen economic analysis shown in the "compelling scenario" is not the point. What is important is that it serves as a basis for constructive criticism—not just dismissal without facts or convoluted arguments that focus on a single aspect of the proposal without attempting to formulate alternative solutions. After all, hydrogen fuel from water is the only long-term answer to the transportation fuel problem, and we will have to deal with it sometime in the future.

At $100+ per barrel, the United States is spending more than $537 billion per year to buy imported oil. In most cases, this oil must be shipped across the ocean to refineries for processing into gasoline and other carbon products, and then delivered by pipeline and trucks to their destinations. According to our preceding analysis, the cost to generate, and deliver to a gas station pump, the equivalent amount of imported oil in the form of liquefied hydrogen fuel is about $430 billion per year (without needing the additional cost of refining and transportation). Imagine what it would mean to our economy if we spent all of the imported oil money in the United States.[*]

---

[*] About 88 percent of crude oil is used in making transportation fuels, but only 53 percent of the crude oil results in transportation fuel after processing. Since much of the refining process requires hydrogen to be extracted from residual oils, an additional amount of crude oil is needed to make the fuels.

The "compelling scenario" described in subchapter 3.1 is for hydrogen fuel equal to about 10 percent of our current United States transportation fuel consumption needs. Therefore, it needs to be repeated ten or more times. We could begin with a pilot program like the one described, but without a long-term commitment to the manufacturing volumes needed to achieve the final objective, we may not meet the mass-produced equipment price targets. And if the price targets are not met, the project may fail. Therefore, a very ambitious initial phase needs to be started.

Accomplishing goals of the magnitude described in this book, using previous studies (that assume existing costs), can be very dangerous, misleading, and shortsighted. What is needed is a vision of the future and a will to get there. When President Kennedy said we will put a man on the moon in this decade, he didn't say let's conduct endless studies to see when and if we should go.

Let's "just do it."

3.3 The *Hindenburg*

So what is it about the *Hindenburg* that causes you so much concern? Haven't you seen gasoline-powered cars and airplanes explode? If so, why do you think the *Hindenburg* is so much more of a problem? Is it because the radio commentator said it was the worst thing that he has ever seen? Or is it because the *Hindenburg* film has been played so many times on television that everyone has etched in their mind that hydrogen must be a horrible thing? (see figure 3.3.1)

Figure 3.3.1. *Hindenburg* **Disaster Photo**

The fact of the matter is, the *Hindenburg* is an example of one of the safety aspects of hydrogen fuel. What the *Hindenburg* showed was that hydrogen is so light that it disperses and floats skyward very quickly. Therefore, when hydrogen burns, it quickly disperses and gives off negligible radiated heat. It is this property of hydrogen that allowed sixty-two of the ninety-seven people on board the *Hindenburg* to survive. Of the remaining thirty-five people, thirty-three died from falling or jumping, while only two people died of burns.

Contrary to what you may have heard, the *Hindenburg* fire was not started by the ignition of hydrogen gas. It was started by a lightning bolt ignition of the iron oxide/aluminum paint covering, which was highly flammable. Proof of this was demonstrated sixty years after the accident using debris from the crash. The chemistry of the paint formula resembled the fuel for a modern-day booster rocket.

You may remember Ben Franklin's famous kiteflying experiment where lightning struck an electric charge collector on a kite that transmitted electricity to the ground. In the case of the *Hindenburg*, it was an 800-foot-long electric charge collector. And when the *Hindenburg* became grounded by dropping its landing lines, the experiment was complete, and the electrical discharge in the *Hindenburg*'s skin started the fire. The *Hindenburg* would have burned and crashed if it had been filled with helium.

As eyewitnesses have noted, the hydrogen fire started considerably after the *Hindenburg*'s surface skin started to burn, and the fire was over in less than a minute. The diesel fuel and other heavier-than-air components of the *Hindenburg* continued burning for hours on the ground.

It should also be remembered that two hundred people in the landing assist team were below the *Hindenburg* holding or reaching for mooring ropes when the *Hindenburg* caught fire. If the *Hindenburg* had carried the same amount of gasoline as the energy released by burning its 7.2 million cubic feet of hydrogen, the loss of life would have included many more of the crew, passengers, and the two-hundred-member landing team.

Perhaps the reader is unaware that the *Hindenburg* had crossed the Atlantic twenty-one times prior to the accident. Its predecessor, the *Graf Zeppelin*, a smaller hydrogen-filled airship, logged over one million miles. It made 650 flights, and over eighteen thousand passengers were delivered safely during the nine years that the *Graf Zeppelin* flew. It flew 144 flights nonstop to and from Berlin, Rio de Janeiro, and New York.

One final point about the *Hindenburg*: the fact that the *Hindenburg caught fire* and did not explode, à la the TWA flight 800,[*] should be viewed as a safety characteristic of hydrogen when compared to gasoline.

---

[*] TWA flight 800 from New York to Paris, in July 1996, went down because of a spark created by wires in the fuel compartment. To solve this problem, current-day aircraft fuel tanks have oxygen extracted from them so that ignition cannot occur. If hydrogen had been used in the TWA fuel tank, it would have burned and not exploded.

To further illustrate the fast-burning, dispersed, and low-heat radiation aspects of burning hydrogen, researchers at Miami University leaked hydrogen from a tank inside an automobile and set it aflame. Measurements inside the vehicle showed no more than 4 °F rise in temperature. If the leak was from a carbon-based fuel like gasoline, the glowing hot-soot particles would have radiated high temperatures to the inside of the automobile, and the gas tank would likely have exploded (hydrogen self-ignites at 1085 °F, and gasoline self-ignites at 934 °F).

Before proceeding further with the safety aspects of hydrogen versus gasoline, please refer to table 3.3.1. As shown, hydrogen is rated the safest when compared to gasoline and methane. In fact, you may be surprised to know that the National Aeronautics and Space Administration (NASA) and Air Products Company (the largest producer of hydrogen in the United States) rate hydrogen safer to use than gasoline in the event of spills and accidents.

Table 3.3.1

	Gasoline	Methane	Hydrogen
Toxicity of Fuel	3	2	1
Toxicity of combustion	3	2	1
Density	3	2	1
Specific Heat	3	2	1
Ignition limit	1	2	3
Ignition energy	2	1	3
Ignition temperature	3	2	1
Flame temperature	3	1	2
Explosion energy	3	2	1
Flame Emissivity	3	2	1
Totals	30	20	16
Safety Factor øs	0.53	0.80	1.00

*1, safest; 2, less safe; 3, least safe

A comparison of various fuel properties

Properties		Hydrogen	Methane	Gasoline
Lower heating value	(kWs/g)	120	50	44.5
Self-ignition temperature	(Degrees C)	585	540	228-501
Flame temperature	(Degrees C)	2045	1875	2200
Ignition limits in air	(Vol. %)	4-75	5.3-15	1.0-7.6
Minimal ignition energy	(mWs)	0.02	0.29	0.24
Rate of flame propagation in air (Stoichiometric composition)	(cm/s)	265	40	40
Detonation limits	(Vol. %)	13-65	6.3-13.5	1.1-3.3
Detonation velocity	(km/s)	1.48-2.15	1.39-1.64	1.4-1.7
Theoretical explosion energy	(kg TNT/m^3 gas)	2.02	7.03	44.22
Diffusion coefficient in air	(cm^2/s)	0.61	0.16	0.05

Re: http://www.ocees.com/textpages/txthydrogen.html

However, some critics believe that hydrogen is too unsafe because it will easily ignite (from static electricity, electric sparks, and open flames) and produce an almost invisible blue flame that will cause people to be burned. While the safety aspects of spark ignition and an "almost invisible flame" should not be minimized, it needs to be put into context. First, there needs to be a leak, a source of oxygen, and a source of ignition for a flame to develop. (Note: In the previous TWA flight 800 footnote, oxygen is currently removed from aircraft fuel compartments to prevent ignition.) Liquefied hydrogen tanks and other handling devices need to have leak-detection equipment, alarms, and shutoff valves. A fiber-optic leak-detection device is shown in figure 3.3.2. This equipment is well understood and applied in current-day practice. Second, the redundancy of double-walled liquid hydrogen tanks and piping, and leak resistant and redundant seals and connections, including the proper selection of materials and quality inspections, make a leak highly improbable. If a leak and subsequent flame should occur, it can be extinguished with a dry powder fire extinguisher.

Figure 3.3.2

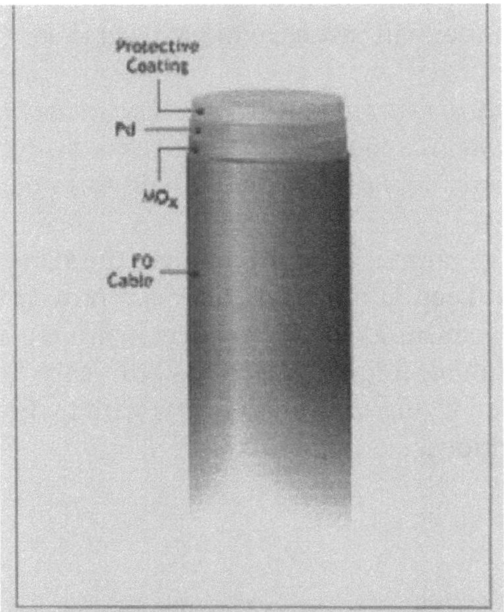

Schematic of fiber optic cable with chemochromic hydrogen sensor deposited on end.

With all of the above being said, mechanics who work with hydrogen equipment need to be certified so that proper precautions are taken. In addition, it should be mentioned that hydrogen-fueled vehicles should not be stored or serviced in an enclosed garage that does not have adequate ventilation and leak-warning devices. It should, however, be pointed out that hydrogen gas will cling to the ceiling of an enclosed garage and try to escape through any available opening. Because of this inherent property of hydrogen, it will generally not cause bodily harm if inadvertently ignited.

One other idea that may be considered is to include an odor-causing agent in the hydrogen fuel to allow it to be sensed without a leak detection device—this approach was used with the *Hindenburg*'s hydrogen gas. While an odor-causing agent may be acceptable for internal combustion engines that run on hydrogen, it may not be acceptable for hydrogen fuel cell vehicles. This is because purity of the hydrogen fuel is important with regard to the fuel cell performance and life expectancy. In this case, any additive in the hydrogen fuel may be detrimental. Also, because hydrogen will quickly disperse into the atmosphere, an odor-causing agent may not be practical. Providing ventilation and leak sensors, like the fiber-optic device mentioned above, may prove to be the best and only prudent approach, and users of hydrogen fuel will just have to get used to it.

Finally on the safety of hydrogen fuel, it was said at the turn of the twentieth century that gasoline was too unsafe to be used in transportation vehicles. And now we are hearing the same thing about hydrogen fuel.

I do not want to minimize or underestimate the dangers involved with hydrogen fuel, but keep in mind that there are proven ways to deal with these dangers. Because, as stated earlier, hydrogen fuel is inevitable sometime in the future, future generations will learn how to live with it. It's time we stopped trying to find problems with hydrogen fuel and start trying to find solutions.

3.4 The Glass Is Half Full

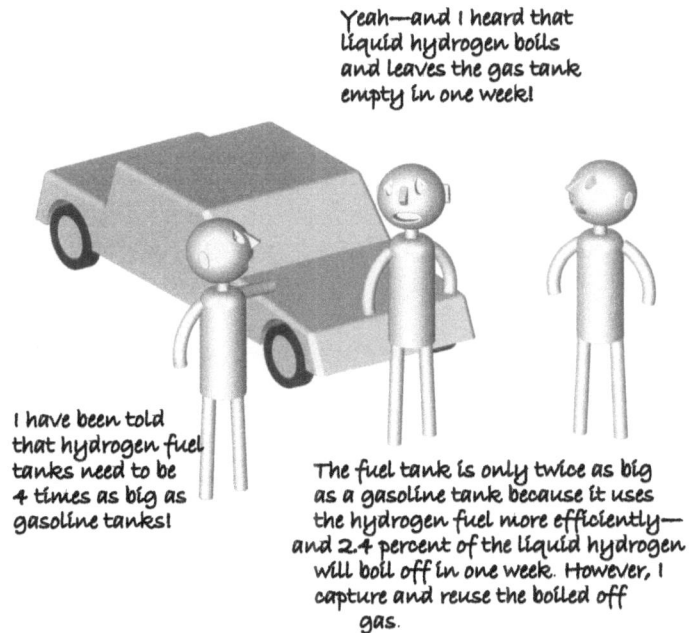

Yeah—and I heard that liquid hydrogen boils and leaves the gas tank empty in one week!

I have been told that hydrogen fuel tanks need to be 4 times as big as gasoline tanks!

The fuel tank is only twice as big as a gasoline tank because it uses the hydrogen fuel more efficiently—and 2.4 percent of the liquid hydrogen will boil off in one week. However, I capture and reuse the boiled off gas.

Are you an optimist or a pessimist? Do you look at half a glass of water as being half full or half empty? Well, I am an optimist, and the glass is half full and not half empty. It is not clear whether the pessimists get it or not. Why find reasons for not using hydrogen fuel rather than finding solutions to make it work? After all, if our generation does not use hydrogen fuel for transportation, the next generation will. And that's a fact that has at least a 90 percent certainty.

Remember, in this book, I am not recommending that hydrogen fuel be used as a stationary power source; instead, I am recommending that it be used only as a portable power source. Electric power generation and home heating are examples of *stationary* power requirements, whereas cars, trucks, boats, airplanes, and trains are examples of *portable* power requirements. If renewable electrical energy—wind and solar—are used in the process of making hydrogen fuel, then they can also be used to produce supplemental electricity for nonintermittent forms of electric power generation—like that produced by natural gas, coal, and nuclear.

With the above being said, let's examine what the pessimists are saying about hydrogen:

1. **Renewable electric energy would be better spent charging battery-powered vehicles.**

Comments

We discussed battery-powered vehicles briefly in subchapters 2.3 and 3.2. As stated, a battery-powered vehicle is generally too heavy, uses too much space for the batteries, and requires a long charge time.

Probably the most advanced battery technology today is lithium ion. For battery size and weight, lithium-ion batteries are about two to three times more effective than nickel metal-hydride batteries (currently used in hybrid vehicles) and four times more effective than lead acid batteries (the kind currently used in your car). Thereby making lithium-ion batteries potentially more capable of realizing the performance and mileage range of gasoline-powered vehicles. *

Currently, research is being done to reduce the space and weight requirements for lithium-ion batteries by more than 50 percent. When and if this is achieved, the battery-powered vehicle may have a place—or niche

---

* Although the cost of lithium-ion batteries may not be an issue in the future, the current replacement cost of batteries, in battery-powered vehicles, ranges from $2,000 for lead-acid batteries to $20,000 for lithium-ion batteries. Today's lithium-ion batteries lose about 20 percent of their charge capability (by self-discharging) in one year at 77°F and have a life expectancy of between twenty-four to thirty-six months from the time of their manufacture.

market—alongside hydrogen fuel cell or hydrogen-hybrid vehicles. But it would appear that the greater range, greater interior space, significantly less refueling time, longer life expectancy, better recycling capability, and better performance of hydrogen-fueled vehicles will always be preferred by most consumers.

If advanced batteries are used in vehicles in conjunction with a hydrocarbon-fueled engine (a hybrid), we will defeat the purpose of eliminating vehicles that use carbon-based fuels. However, using batteries in conjunction with a small hydrogen-fueled engine might work well as a hybrid. More on this later.

It takes about 78 kilowatt-hours of electrical energy to produce and deliver 36 kilowatt-hours of liquefied hydrogen fuel energy to your fuel tank. About 55 percent of this hydrogen fuel energy is then lost in transforming the hydrogen fuel back into electrical energy via a fuel cell in your car (i.e., 20 kilowatt-hours remain to power the motors that propel the car—or 25 percent of the original 78 kilowatt-hours). In contrast, if we charge 80 percent efficient batteries with 78 kilowatt-hours of energy, we would have 62 kilowatt-hours of energy available from the charged batteries to drive the motors that propel the car. At first glance, you may be saying that this sounds like a good idea, but so does cutting out the batteries completely and attaching an electric cord to power the car. The question is one of practicality when a better alternative exists.

Since, as a rule of thumb, we lose 10 percent of the vehicles mileage range for every 10 percent of added weight, it is quite likely that most of the improved energy advantage—of the batteries mentioned above—will be lost. As a result, you will be left with a car that has much less range, much less interior space, significantly more refueling time, less life expectancy, significantly greater cost of ownership (due to battery replacement), less recycling capability, and sluggish performance when compared to hydrogen fuel cell and hydrogen-hybrid vehicles. However, with that being said, a plug-in hybrid may have some promise—more on this later.

Turning vehicle wheels with electric motors has well-known advantages of torque, ruggedness, reliability, simplicity, controllability, quietness, and low cost. The battery hybrid, and fuel cell-powered cars use this technology to their benefit. Fortunately, battery-powered and gasoline-powered hybrid vehicles have helped lay the groundwork that will speed up the transition to hydrogen-powered hybrids and hydrogen fuel cell vehicles.

2. Hydrogen fuel costs too much and is made from hydrocarbon fuels that pollute the atmosphere—where is the advantage?

Comment

Hydrogen is currently made from hydrocarbons (like coal, natural gas, methane, and oil) because this is currently the least expensive method. As a result, we will continue to generate air pollution and greenhouse gases. Here again, as described in this book, we can economically make hydrogen from water using limitless "green" renewable wind and solar energy and water.

3. Another hydrogen concern is that it would take four liquefied hydrogen trucks to deliver the same amount of hydrogen fuel energy to gas stations as one gasoline delivery truck (due to the nature of liquefied hydrogen, it occupies about four times as much volume as an equivalent volume of gasoline).

Comments

Based upon this criticism, it means that more fuel is consumed in making deliveries, and more trucks would be needed—a problem especially in traffic-congested areas. This is a paradigm that suggests that trucks deliver liquefied hydrogen fuel to gas stations. The simplest way to deliver liquefied hydrogen to gas stations is to deliver low-cost electricity and water to the gas station and let them make their own fuel, thereby eliminating truck delivery.

Although some gas stations may make their own hydrogen fuel in the future, our "compelling scenario" still relies on trucks. But the trucks are double tractor-trailers, and the fuel tanks are "dropped off" at the gas station (because of their lighter weight compared to equivalent tanks filled with gasoline). The empty tanks are then transported back to a nearby fuel-dispensing station for refilling (sort of like propane gas is dispensed for backyard barbecue grills).

Two tanks carrying 2,500 kilograms of liquefied hydrogen can be delivered per trip to a gas station each day. And because the centralized hydrogen fuel processing stations are located nearby, the number of double tractor-trailers and resulting transportation fuel consumption may be reduced compared to some instances of longer-distance gasoline delivery. Also, when fuel-efficient hydrogen fuel cell and hybrid vehicles are in widespread use, the amount of hydrogen dispensed at each gas station will be dramatically reduced. Additionally, the hydrogen fuel used to transport the hydrogen fuel will be more fuel-efficient than gasoline.

4. **There is another critic who stated that a 747 aircraft would have to be three times as big to accommodate pressurized hydrogen fuel tanks.**

Comments

First of all, liquefied hydrogen fuel is more compact and lighter weight in its container than pressurized hydrogen gas in its container. Second, future aircraft will probably need to be redesigned for hydrogen fuel—probably with the fuel tank on top of the aircraft.

There is one proposal that places the liquefied hydrogen fuel tank in the body (fuselage) of an aircraft between the cockpit and the passenger compartment. Such a placement would make it impossible for a terrorist to gain access to the cockpit.

Liquefied hydrogen fuel is much more efficient for jet aircraft and is 70 percent lighter in weight (specialized large-size lightweight aluminum or titanium hydrogen fuel tanks would probably reduce this weight advantage to about 30 percent to 40 percent). Because of these considerations, aircraft designers conducted a study that showed that the fuel flight efficiency could roughly double, thereby requiring an approximately two times larger hydrogen fuel tank than its comparable jet fuel tank. Their studies also showed improved safety characteristics, smaller wing areas, shorter runways, quieter operation, and—of course—less pollution.

5. **It has been stated that hydrogen is made from water, and thus the U.S. freshwater consumption rate would go up 10 percent in an all hydrogen fuel economy.**

Comment

This criticism is not true because it currently takes eighteen gallons of water to make one gallon of gasoline from crude oil, whereas it takes about 2.5 gallons of water to make an equivalent gallon of hydrogen (in terms of energy content). Therefore, the U.S. freshwater consumption rate will go down more than 10 percent rather than up 10 percent. And keep in mind that when hydrogen is burned, its by-product is water that returns to the atmosphere. This is almost a perfect cycle when you consider that it combines with the oxygen that is released into the atmosphere during the production of the hydrogen gas.

6. **Some critics claim that if you let a liquefied hydrogen-filled automobile sit for a week, the fuel tank would be empty due to "boil off."**

Comments

This criticism is also not true. A liquefied hydrogen-filled automotive fuel tank loses about 1 percent to 3 percent of hydrogen fuel per day, and that percentage is highly dependent on the storage volume to tank surface area ratio and on the insulation quality of the tank design. Therefore, at the midvalue of 2 percent, it would take nearly two months for a 10-kilogram liquid hydrogen fuel tank to become 70 percent emptied (10 kilograms would become 3 kilograms).

"Boil off" can be an issue if you do not drive your automobile very much and have infrequent refills. In this case, pressurized hydrogen gas may be a better solution. For this reason, hydrogen fuel gas stations should probably offer both gaseous and liquefied fuel. However, if you fill your tank once a week, there will be about a three-day dormant period after each refill before "boil off" begins. This means that if you have a 10-kilogram tank and a 2 percent "boil off" rate for four days (seven days minus three days dormancy), you will be losing about 0.8 kilograms of hydrogen fuel during the first week. If you think of this as 42 kilograms per year, at $4.63 per kilogram, the loss would be equal to $195 per year. This loss then needs to be balanced against the gasoline savings that would result from using hydrogen fuel ($5.92 to $33.70 per 20-gallon fill-up as shown in subchapter 3.1).

Further to the "boil off" issue, it should be mentioned that studies are currently being conducted to improve the "boil off" dormancy period from three days to about twelve days through "active cooling" systems (the Linde "CooLH2" system will be discussed later). Additionally, other researchers are trying to find cost-effective ways of capturing the "boil off" gas by using metal hydrides, zeolites, and carbon nanotubes. Thus making the "boiled-off" hydrogen gas available for use as a fuel in the operation of the automobile (e.g., a small fuel cell) or as a gas station rebate when filling the tank.

The development of other types of hydrogen fuel tanks will be discussed in a later subchapter. Since there are so many promising possibilities, only time will tell the outcome.

7. Some other critics claim that we don't have enough experience using hydrogen-to-fuel automobiles and trucks, and therefore, we need to proceed slowly and cautiously.

Comments

What is generally not known is that engineers in Germany and England had been using hydrogen as an automotive fuel since the early 1900s. By the year 1930, literally thousands of hydrogen-fueled vehicles were in operation in the two countries. At that time, a German engineer named Rudolf Erren developed a fuel injection system that eliminated the carburetor that was poorly suited to inject a gaseous fuel. Since the remaining parts of the internal combustion engine were unchanged, a conversion cost was relatively small, and the vehicle was able to operate on either hydrogen or hydrocarbon fuels with a flip of a switch. All major engines at that time were modified including those manufactured by MAN, Daimler-Benz, and Beardmore.

In the 1980s, a self-service liquid hydrogen pump was used by Los Alamos National Laboratory investigators to refuel a modified 1979 Buick. After using the system for over a year, the investigators concluded that "liquid hydrogen storage and refueling of a vehicle can be accomplished over an extended period of time without any major difficulty."

A BMW V12 engine was able to operate on gasoline or liquid hydrogen with the flip of a switch from inside the vehicle. BMW engineers reported that "if liquid hydrogen fuel storage systems are used, which are similar to gasoline, that drivers would not have to give up performance, vehicle size, or range."

8. **Researchers at Caltech have claimed that release of large quantities of hydrogen into the atmosphere (leakage of 20 percent of the world hydrogen fuel economy usage) could cause a slowing of the stratospheric ozone levels as chlorofluorocarbon emissions are reduced.**

Comments

First of all, the release of hydrogen into the atmosphere would be mostly a result of liquefied hydrogen "boil off." "Boil off," as discussed above, can and should be controlled for economic reasons. Therefore, 20 percent leakage is an unrealistic assumption.

Second, not to dispute the Caltech researchers, but there are many unanswered questions related to their study, such as the fact that molecular hydrogen may never reach the stratosphere because the earth absorbs molecular hydrogen like a sponge.

9. **Because the U.S. car fleet takes roughly fourteen years to turn over, little can be done to change car technology in the short term.**

Comments

First, the fourteen-year cycle could be improved by retrofitting current vehicles with dual fuel tanks—one for gasoline and one for hydrogen. Some vehicle owners will want to do this if the retrofit kit is cheap enough, and there is a government incentive to do so.

Another approach would be to provide car owners with an incentive to trade in their gas-guzzling car for one that is less polluting. The incentive could be a rebate based upon the difference between the new and old car's gas efficiency. In this case, the gas-guzzler would be scrapped, since society would consider the old car worth more to them dead than alive. Also, as a result, Detroit would sell more cars that would yield benefits in terms of more jobs and an improved economy.

10. It has been postulated that the hydrogen transition would require a huge federal cash outlay (say $100-$300 billion) for a crash program along the lines of the Apollo Moon Program.

Comments

This postulation assumes that government would put up most of the money, when in fact there is a huge economic incentive for private enterprise to put up most of the money.

The early stages of the program will be directed toward conducting feasibility tests to confirm that the recommended approaches are correct. These early stage feasibility studies are relatively inexpensive and could be funded by rearranging existing government-sponsored programs. A word of caution should be given here. It is imperative that the feasibility programs proceed with preestablished milestones that prevent researchers from making a career out of their studies.

The current five-year United States Government-funded hydrogen program is funded at a level of $1.7 billion; this is considerably less than the stated $100-$300 billion, but it is also considerably less than what is required. Private industry has already spent billions of dollars to provide various pieces to the puzzle.

3.5 Fill'er Up

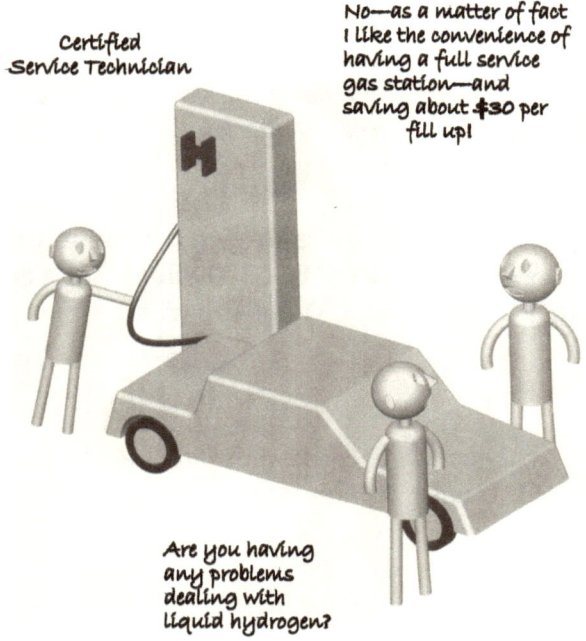

As you already know, I am recommending that, in the initial stages of building a hydrogen infrastructure, we use liquefied hydrogen. But beyond that, we can expect gaseous hydrogen to be used. We can also expect that, someday, electricity/water or gaseous hydrogen will be delivered to each gas station, rather than by the tractor-trailer system previously described.

Liquefied hydrogen may sound scary to some, but keep in mind that certified gas station technicians will be dispensing the fuel. And for the most part, the process of refueling will be no more difficult than current practice. Only in this case, you will not have to get out of your car.

The liquefied hydrogen fuel tank will be considerably different from the one that you currently have. It will be somewhat bigger than before and will be comprised of a double-walled tank that is similar to a thermos bottle. Since the liquefied hydrogen will be extremely cold, the tank will be much better insulated than a conventional thermos bottle and will not cause frost to form on the outside surface. It will be more like the stainless steel liquid nitrogen bottle used by dermatologists. If you have any experience with dermatologists, you might have noticed that the liquid nitrogen bottle is handled with bare hands, even though the liquid nitrogen inside is at an extremely low temperature. It is so cold that, in a science lab experiment, an air-filled balloon can be turned into a glasslike shell when dipped into it. In this experiment, the lab instructor will drop the glasslike balloon onto a hard surface and cause it to shatter like a lightbulb.

Before I discuss the specific design of the liquefied hydrogen fuel tank, let's first recall the main advantage of a hydrogen automobile. As you may remember, if you needed twenty gallons of gasoline to fill your gasoline-powered automobile, you would need only ten equivalent gallons (kilograms) of liquefied hydrogen fuel for the same range and performance with your hydrogen fuel cell car. And as previously stated, if hydrogen fuel costs almost the same as gasoline, and you paid $80 for the gasoline (at $4.00 per gallon), you would only pay $46.30 (at $4.63 per kilogram) for the hydrogen fuel.

Hydrogen Fuel Cell Cars

You have all probably heard by now that hydrogen fuel cell cars are too expensive and don't last very long. This is one of the arguments that have been made by critics of the hydrogen fuel scenario. If this were true, then why have almost all the automobile manufactures in the world developed fuel cell cars? Do you think it is because they just like to work on exotic new automobile designs or have so much extra money that they just enjoy investing in something that has no hope of success?

The fact is that automotive manufactures know, beyond much doubt, that hydrogen-fueled vehicles are going to be used in the future. In fact, the gasoline-powered hybrid of today could be the hydrogen-powered hybrid of tomorrow. It lays the groundwork for electric motor-driven cars. And as stated before, both the hybrid and the fuel cell vehicles will use at least one electric motor and drive by wire technology (to be discussed later). One version of the hybrid uses a small gasoline-fueled internal combustion engine to charge batteries and power the vehicle. When additional power is required for acceleration, a battery-powered electric motor kicks in (when braking, the electric motor becomes an alternator to assist in the battery charging process). On the other hand, the fuel cell converts hydrogen fuel directly into electricity that powers the electric motor drive system.

To prove that the major automobile manufacturers have developed, and are continuing to develop hydrogen fuel cell automobiles, you only need to look at the tremendous investments that they are making. For example, it has been disclosed that at one time, six hundred people were working on fuel cell technology at General Motors.

To put this into better perspective, the photographs shown in figure 3.5.1 below illustrate this point:

Figure 3.5.1. Fuel Cell Automobiles

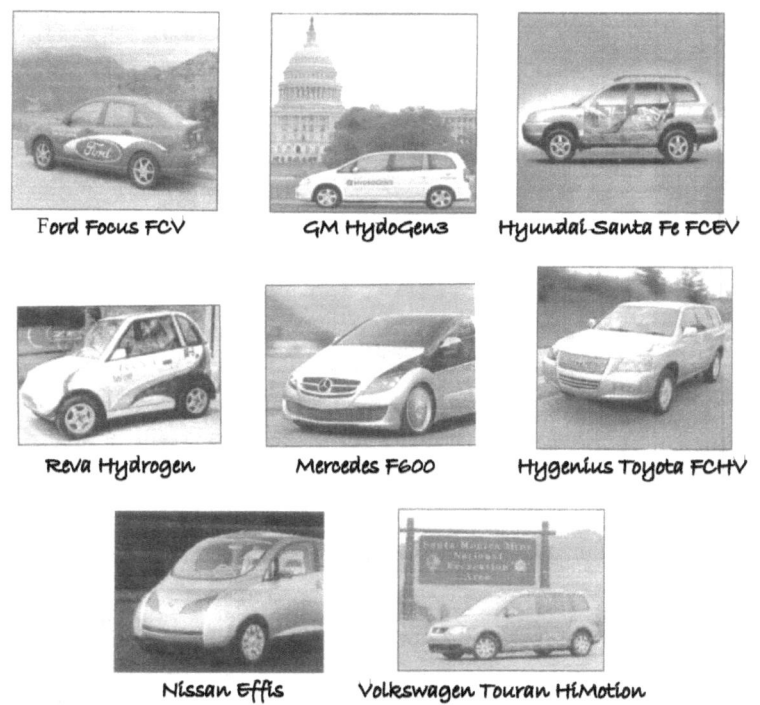

Ford Focus FCV   GM HydoGen3   Hyundai Santa Fe FCEV

Reva Hydrogen   Mercedes F600   Hygenius Toyota FCHV

Nissan Effis   Volkswagen Touran HiMotion

As an example of one of these vehicles, the GM HydroGen 3 specifications are as follows:

    4.6 kilograms of liquefied hydrogen
    54.3 miles per kilogram of hydrogen
    250 miles range

60 kw/82 hp
3500 pounds weight
Seating for 5
Full-size trunk
99 mph top speed
0 to 99 mph in less than 16 seconds

The Skateboard Chassis

In addition, the HydroGen 3 and other fuel cell automobiles take advantage of having a flat "*skateboard*" chassis. This chassis, shown in figure 3.5.2, uses a compact "drive by wire" technology for steering, braking, and throttling that permits greater freedom for designers to configure the upper body, which means that the designer does not have to contend with bulky engine compartments, an awkward center cabin hump, or conventional steering wheel. This novel approach also allows bodies to be interchangeable. For example, an owner could have new personalized bodies "plugged in" to their used chassis at the dealership, or do it themselves—turning, say, a family sedan into a luxury car.

Figure 3.5.2. The Skateboard Chassis

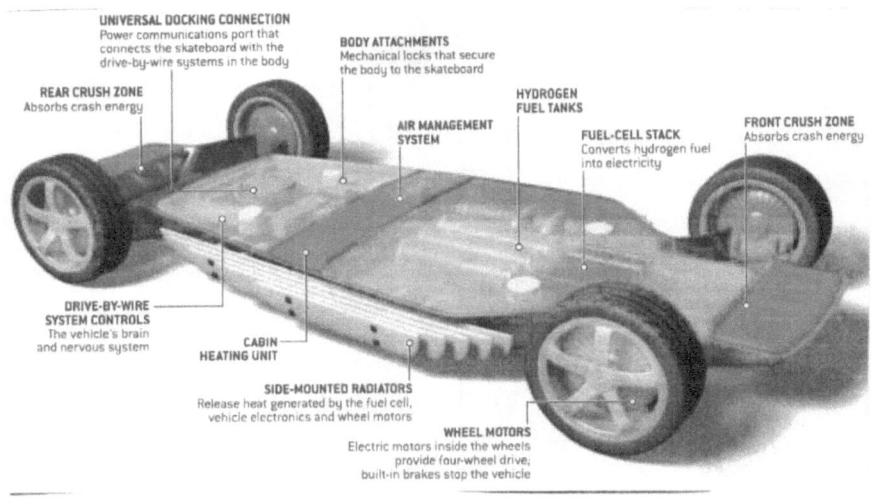

Hydrogen Fuel Cell Technology

A hydrogen fuel cell is the key component of the above vehicles. However, it has been in development for a long time and still has some technical problems that need to be resolved. In addition, mainly because it is currently not mass-produced, its cost is very high.

A leading developer of hydrogen fuel cells is the Ballard Company, which is currently partnered with many of the previously mentioned automobile manufacturers. In recent years, the fuel cell's cost has dropped dramatically from $3,000 per kilowatt to $60 per kilowatt, based upon volumes of 500,000 fuel cells per year. This cost is expected to drop to $30 per kilowatt by the year 2010, which is equal to the current cost of an internal combustion engine.

The hydrogen fuel cell technology is based upon what is called a proton exchange membrane (PEM) as illustrated in figure 3.5.3.

Figure 3.5.3. Proton Exchange Membrane (PEM) Fuel Cell

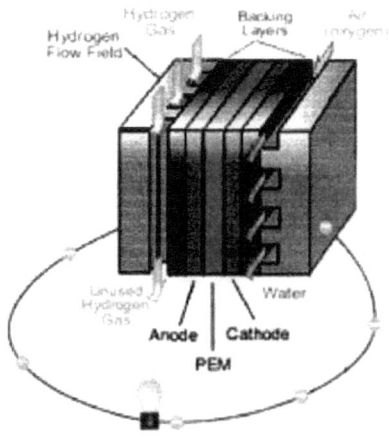

As shown, hydrogen gas is forced through an anode, proton exchange membrane, and cathode, where it mixes with air to form water. The resulting process releases electrons to produce electricity.

Aside from cost, the hydrogen fuel cell has several technical problems that Ballard is in the process of resolving. These problems are power density, freeze point, and life expectancy. In Ballard's 2010 plan, the power density is expected to increase from 1.8 kilowatts per liter of hydrogen to 2.5 kilowatts per liter, and the freeze point is expected to increase from -13 °F to -22 °F. At the same time, Ballard expects the durability to improve from 3,200 hours to 5,000 hours (5,000 hours at 50 miles per hour equals 250,000 miles).

To improve cost, Ballard is reducing the thickness of the platinum used to coat the electrodes from 0.7 grams per kilowatt to 0.2 grams per kilowatt. A previous thickness reduction from 1.0 to 0.7 grams per kilowatt did not have any effect on durability or performance.

Since platinum is the most precious of precious metals (at more than $1,000 per ounce), it becomes a key factor in the future success of fuel cells. Using the above-mentioned 60-kilowatt General Motors "HydroGen 3" concept car, as an example, we can see that the platinum alone—at 0.2 grams per kilowatt—would cost $420 (assuming $1,000 per once). Since the 60-kilowatt fuel cell stack would cost $1,800, the platinum would equal 23 percent of its cost (the platinum-coated anodes and cathodes are recyclable and could limit the total amount of platinum that would be consumed by future vehicles).

Therefore, I recommend that in conjunction with developing a hydrogen fuel economy, *the United States Government should invest heavily in platinum as a strategic metal* and discontinue its current use as jewelry. Additionally, since almost 10 tons of ore must be mined to extract only 31 grams of platinum, it would make sense to invest in more economic methods of ore extraction and processing.

With the above obstacles yet to be overcome, the United States needs to develop a hydrogen-consuming infrastructure that is based upon today's technology. Since it will normally take up to fourteen years to clear out today's inventory of conventional gasoline-powered cars, we can help this process along by retrofitting existing cars with hydrogen fuel technology. *Here is one way to do it:*

> Provide incentives to equip about twenty thousand gas stations with liquefied hydrogen fueling capabilities.

> During this same time frame, provide incentives for automobile manufacturers to sell conversion kits for previous-year vehicles

to run on gasoline or liquefied hydrogen—the "combo" vehicle.[*]

Another version of the above "combo" vehicle, is to remove the gasoline tank and replace it with a 5 kilogram hydrogen tank. This conversion would be less costly and would work for those people with second cars that are used primarily for short distance driving.

Provide manufactures with incentives to produce new vehicles that are "combo" conversion-ready—preferably hybrids that could have smaller fuel tanks due to their improved efficiency.

Provide incentives to repair shops to set up facilities to retrofit vehicles at a low cost.

Provide incentives to selected volunteer vehicle owners who convert to "combo" vehicles.

There are various ways to pay for the above incentives, one of which could be an added tax on the price of the hydrogen fuel. In this case, the tax would equal the gas mileage savings that would result from using hydrogen—assuming the hydrogen fuel price is equal to or better than the

---

[*]   In this instance, the range of the vehicle, which has two fuel supplies, could be optimized, and lighter materials—aluminum or carbon composites versus steel gasoline tanks—could be employed. It is recommended that the gasoline tank hold up to fifteen gallons, and the hydrogen tank hold up to 5 kilograms (the hydrogen tank would be less than two times the volume of the gasoline tank). The gasoline tank would then provide a range of about 300 miles at 20 miles per gallon. The hydrogen tank would provide a 125-mile range since hydrogen is about 25 percent more efficient than gasoline. To provide room for the hydrogen fuel tank, consideration could be given to eliminating the spare tire and jack system. In this instance, flat proof tires—like Michelin's "tweel"—could be used. Removal of these items would also compensate for the increased fuel tank weight, which has the effect of reducing gas mileage.

price of gasoline. In any event, if we can subsidize corn ethanol and wind turbines, we can subsidize the use of hydrogen.

The Hydrogen Hybrid

Before going on, it is worth mentioning a *hybrid hydrogen-powered automobile developed by the Los Alamos National Laboratory in 1995.* The automobile utilized a small 40-kilowatt engine that powered a generator, which charged batteries that would power electric motors to propel the car. The engine ran only when needed to charge the storage system at an optimized speed for maximum efficiency. The five-passenger 2,508-pound automobile accelerated from 0 to 60 miles per hour in 8 seconds *and required 3.75 kilograms of hydrogen for a 300-mile range. This results in a gasoline equivalent mileage of 80 miles per gallon—which is about the efficiency of a fuel cell vehicle.*

The only significant emissions from the vehicle were water vapor and some small amounts of nitrous oxides—less than one-tenth of the California ultralow emission standard. No further comment on this, except to say, that if this type of vehicle were to be developed, it could *provide significant competition for the fuel cell-powered vehicle.*

The Hydrogen Fuel Tank

Many consider the vehicle fuel tank to be the "Holy Grail" in the hydrogen fuel controversy. And as such, it has become one of the biggest roadblocks to moving hydrogen forward as a legitimate fuel source. In other words, it is thought that a breakthrough technology is needed.

Since currently available liquid hydrogen fuel tanks can fulfill all the requirements needed, it is my opinion that a breakthrough technology is not required. If a new and better method is developed, the industry will adapt. In the meantime, considering all of the benefits that will be derived from converting to hydrogen fuel, there is no excuse for not proceeding with the liquefied hydrogen fuel infrastructure now.

Current Hydrogen Fuel Tank Technology

What you are about to read is probably more technical than most readers want to endure. However, because of the controversy regarding the fuel tank, I am going to spend some time, in this subchapter, explaining to the reader why liquefied hydrogen fuel tanks can be used now and do not require a breakthrough technology. This is important because you will need to know how to answer the critics when you are confronted with this issue.

To give the reader an idea of what is currently available in liquefied hydrogen fuel tanks, please refer to the Linde tank illustrated in figure 3.5.4

As you can see, this fuel tank is much more complex than a conventional gasoline fuel tank. For instance, there are safety devices, heat exchangers, and "boil off" systems that gasoline tanks do not need. However, this fuel tank is currently available, it works, and its functions are—for the most part—*transparent to the vehicle owner.* To make the tanks serviceable, the water pump, heat exchangers, and solenoid valves need to be easily accessible, and the tank needs to be mounted where it can be replaced if necessary. Replacement of a noncorroding tank would, however, be very unlikely during the life of the vehicle.

Figure 3.5.4. Linde—Liquefied Hydrogen Fuel Tank

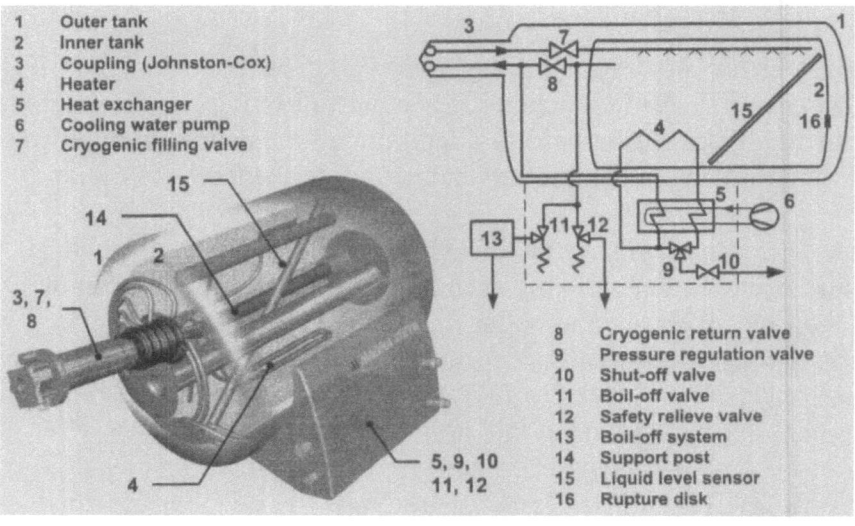

1	Outer tank
2	Inner tank
3	Coupling (Johnston-Cox)
4	Heater
5	Heat exchanger
6	Cooling water pump
7	Cryogenic filling valve
8	Cryogenic return valve
9	Pressure regulation valve
10	Shut-off valve
11	Boil-off valve
12	Safety relieve valve
13	Boil-off system
14	Support post
15	Liquid level sensor
16	Rupture disk

The design utilizes an inner and outer tank made of one-eighth-inch-thick stainless steel. The space between the inner and outer tanks is filled with forty layers of superinsulation with a weight-to-area ratio of 0.3 to 0.6 pounds per square foot. The superinsulation is comprised of aluminized polymer foils separated by glass fiber spacers. Then a vacuum is applied to the space to reduce thermal convection. The support structures keeping the inner tank in position relative to the outer tank are made of glass or carbon-reinforced plastics. If the vehicle is not used for more than three days, the heat entry leads to a "boil off" rate of 1 percent to 3 percent per day.

During the filling procedure, both the cryogenic filling valve (7) and the cryogenic return valve (8) are opened. Liquefied hydrogen flows from the filling station via a Johnson-Cox coupling (3) and the filling valve into the inner vessel (1). In order to keep the inner tank pressure low, evaporated gaseous hydrogen leaves the inner tank via the cryogenic return valve and flows back to the filling station. After finishing the filling procedure, both cryogenic valves will be closed. For hydrogen extraction, the cryogenic filling valve remains closed while the cryogenic return valve is open. Gaseous hydrogen leaves the inner tank to the cooling water heat exchanger (5). Hydrogen heats up above ambient temperature and flows farther into the pressure regulation valve (9). If the inlet pressure is above the defined set pressure of the pressure regulator, the partial flow inlet will be closed and no hydrogen can pass through the tank heater (4). Therefore, no additional heat will be led to the inner tank heater and the pressure will decrease. During standby, both cryogenic valves are closed. During long-term parking, the hydrogen pressure in the inner tank rises until the "boil off" valve (11) will limit the "boil off" pressure. Overpressure in the inner tank must not open the cryogenic valves. In case of a fault of the "boil off," the pressure in the inner tank rises until the safety relief valve (12) opens. The last device that protects the tank is the rupture disc (16), which is needed in case of a default of the safety relief valve.

The problems with this tank are (1) excessive weight—22 pounds of liquefied hydrogen to 330 pounds of tank weight; (2) awkward shape; and (3) "boil off" after three days. However, one company in Austria—Magna Steyr—is currently addressing these problems:

1. Tank weight

First of all, a high tank weight to hydrogen weight is a characteristic of small volume tanks because of a high surface area to volume ratio (i.e., the heat transfer surface area is large relative to the volume of liquid contained). As such, the tank weight to hydrogen weight ratio can be dramatically reduced with larger liquid hydrogen volumes; however, because small volume tanks are a requirement, the solution needs to be in the selection of construction materials and more efficient and redesigned components (e.g., insulation material, valves, heat exchangers, pump, tubing, supports, and flanges). It seems obvious to me that aluminum can replace the inner and outer tanks. To give an example, the Linde tank is made of stainless steel that weighs about 210 pounds, while an aluminum tank—adjusted for strength—would weigh about 88 pounds—*a weight savings of 122 pounds.*

2. *Awkward shape*

It is very desirable, for most vehicles, to have a "free-form" hydrogen fuel tank because of its larger size relative to gasoline fuel tanks. Fortunately, liquefied hydrogen requires low-pressure tanks that can be free-formed much more easily than high-pressure tanks. The typical liquefied hydrogen tank design pressure is 133 pounds per square inch (psi) compared to pressurized tanks of 5,000 psi and 10,000 psi. Magna Steyr is evaluating the use of fiber-composite materials that when appropriate orientation of the winding fibers is used, greater strength can be achieved than with stainless steel.

3. *Boil off*

The Linde Company has already designed and tested a liquefied hydrogen tank—similar to the one described herein—that utilizes active cooling to delay the onset of "boil off" for as much as twelve days. Their concept is called "CooLH2." If this proves to be impractical, absorbing materials such as carbon nanotubes, metal hydrides, or zeolite could be utilized to capture the "boil off" for reuse or for credit at the gas station.

Gaseous Hydrogen Fuel Tank Technology

As mentioned in the opening paragraph of this subchapter, the future belongs to gaseous hydrogen tanks, but some technological breakthroughs *are* needed. At present, there are *three currently available contenders to replace liquefied hydrogen fuel tanks* with gaseous hydrogen fuel tanks. They are (1) 1,500 psi metal hydride tanks, (2) 5,000 psi pressurized tanks, and (3) 10,000 psi pressurized tanks. The following describes each approach and how it compares with the currently available Linde liquefied hydrogen fuel tank described above:

Metal Hydride

The 1,500 psi metal hydride tanks (designed by the Ovionics Company in conjunction with Texaco) are available with a full tank weight to hydrogen content weight ratio of about 45 to 1 (15 to 1 for the Linde liquefied hydrogen tank). The ratio of pressurized and absorbed hydrogen gas energy to tank volume is about 1.95 to 1 (2 to 1 for the Linde liquefied hydrogen tank). When applying the metal hydride technology to a fuel cell automobile—like the above-mentioned HydroGen 3—the tank may only need 5 kilograms of hydrogen fuel to provide a good mileage range. This would then result in adding 495 pounds but subtracting about 165 pounds for the liquefied hydrogen tank—a net weight increase of about 330 pounds. This is a vehicle weight increase of about 10 percent, and a rule of thumb is that a weight increase of 10 percent usually results in a 10 percent fuel mileage decrease. There is another downside to metal hydrides, and that is fuel tank price, which is currently about two to three times more expensive than the comparable liquefied hydrogen fuel tank. Also, it is not clear how long the metal hydride will last or how long a refill of pressurized hydrogen gas will be when compared to filling with liquefied hydrogen—a current estimate to fill a 22-pound liquefied hydrogen tank is about two minutes.

5,000 PSI Pressurized Tank

The 5,000 psi pressurized fuel tanks have a hydrogen gas content weight ratio of about 19 to 1 (15 to 1 for the Linde liquefied hydrogen tank). This tank has very little prospect for weight reduction, since the tanks are constructed with an aluminum inner lining and a fiberglass outer shell. The ratio of pressurized gas energy to tank volume is about 0.5 to 1 (2 to 1 for the Linde liquefied hydrogen tank).

10,000 PSI Pressurized Tank

The 10,000 psi pressurized fuel tanks have a hydrogen gas content weight ratio of about 22 to 1 (15 to 1 for the Linde liquefied hydrogen tank). However, the ratio of pressurized gas energy to tank volume is about 0.8 to 1 (2 to 1 for the Linde liquefied hydrogen tank).

All things considered, liquefied hydrogen looks very favorable when compared to the currently available gaseous hydrogen alternatives. In addition, as stated earlier, on tractor-trailer deliveries to gas stations, liquefied hydrogen may still be the best way to go. In fact, even if a better fuel tank were developed, to replace liquefied hydrogen fuel tanks, it would still make sense to deliver liquefied hydrogen to gas stations. Unless of course every gas station had its own hydrogen gas pipeline—a formidable task that may occur in the future but not during the early stages of infrastructure development.

Advanced Hydrogen Fuel Tank Technology

As shown, each of the above fuel tank options has a penalty when compared to liquefied hydrogen, except the Ovionics metal hydride tank, which can contain almost as much hydrogen energy per unit volume. However, the metal hydride has a very large weight penalty that can affect the performance and fuel mileage of the vehicle (990 pounds—to contain 22 pounds of hydrogen fuel).

Meanwhile the Department of Energy (DOE) has a $150-million program to find a solution to hydrogen gas storage. In 2004, thirty universities, ten companies, and ten national laboratories were given contracts to find solutions in a program called the "Grand Challenge." The technologies include

> hydrogen gas tanks,
> liquefied hydrogen tanks,
> metal hydrides,
> carbon-based materials,
> high surface area sorbents, and
> chemical storage.

All of the above methods are considered to be "on board" storage, except for the chemical storage that must be regenerated "off board."

The planned 2005 research and development work included the following:

Hydrogen gas tanks—reduce cost of 350 and 700 bar tanks

Liquefied hydrogen tanks—no reported focus or objectives

Metal hydrides—develop advanced metal hydride materials including light element advanced complex hydrides, destabilized binary hydrides, intermetallic hydrides, modified lithium amides, and other onboard reversible hydrides.

Carbon-based materials—focus on breakthrough concepts of hybrid carbon nanotubes designed to absorb hydrogen gas inside various tube geometries made of "fullereen" carbon particles

High-surface area sorbents—focus on high surface area sorbents such as aerogels, nanofibers, metal-organics, and conducting polymers

Chemical storage—focuses on three "tiers" of "off board" chemical storage: borohydride-water, novel boron chemistry, and innovation beyond boron

Sorry for the technical jargon, but it is shown here to give the reader some idea of what is happening today to resolve the hydrogen fuel tank issue. I will not go into the details for each of these programs, except to say that it does not appear that a great deal of effort is being spent on liquefied hydrogen or on lightweight sorbents like zeolite—which is low-cost and weighs only about 60 pounds per cubic foot—similar to the weight as gasoline.

To give the reader further information about the DOE fuel tank program, the target goals are as follows (note: I will use the metric system to be consistent with the DOE reporting):

Achieve a ratio of contained hydrogen energy to liter of tank volume of 1.5 by the year 2010, and 2.7 by the year 2015 (1 liter = 0.264 U.S. gallons).

Achieve a ratio of contained hydrogen energy to tank weight of 2.0 by the year 2010, and 3.0 by the year 2015.

Achieve a ratio of U.S. dollars to contained hydrogen energy of $4 by the year 2010, and $2 by the year 2015.

The 10-kilogram (22-pound) Linde liquefied hydrogen fuel tank is compared with these objectives as follows:

The 2.0 kilowatt-hours of hydrogen energy per liter of tank volume versus the 2015 objective of 2.7.

The 2.4 kilowatt-hours of hydrogen energy per kilogram of tank weight versus the 2015 objective of 3.0 kilowatt-hours per kilogram of tank weight can be achieved with my improved aluminum Linde tank discussed above.

The U.S. dollars per contained hydrogen energy is unknown by me at this time. However, at an improved Linde tank weight of 70 kilograms of aluminum plus other previously mentioned parts and materials at an average mass-produced cost of $10 per kilogram, the tank would cost about $700. Therefore, for 10 kilograms of contained liquefied hydrogen (360 kilowatt-hours of energy), the cost would be $1.94/kilowatt-hour versus the 2015 DOE objective of $2.00.

Except for the kilowatt-hours of hydrogen energy per liter ratio being 2.0 versus the 2015 target of 2.7 kilowatt-hours per liter, liquefied hydrogen compares very favorably with the 2015 DOE target goals.

If the above technologies ever meet or exceed the target goals, there may be a shift toward using only gaseous hydrogen. If, and when, that happens, the liquefied hydrogen infrastructure will adapt. In the meantime, we will have developed a hydrogen fuel infrastructure.

As previously mentioned, there is considerable experience with safely using liquefied hydrogen. But dispensing of liquefied hydrogen at the gas station pump will probably require qualified and certified personnel. In other words, we will probably have our gas tanks filled the old-fashioned way—with a service attendant. As an additional safety measure, there may be a requirement for users of hydrogen vehicles to be qualified in their use. One way to do this would be take a short test and have an H *stamped on their driver's license.* The main reason for doing this would be to assure that drivers understood basic safety precautions like the need to have their home garage adequately ventilated. Since hydrogen will quickly leak out of an enclosed space, most home garages and service facilities will not require modification.

3.6 Homeland Security

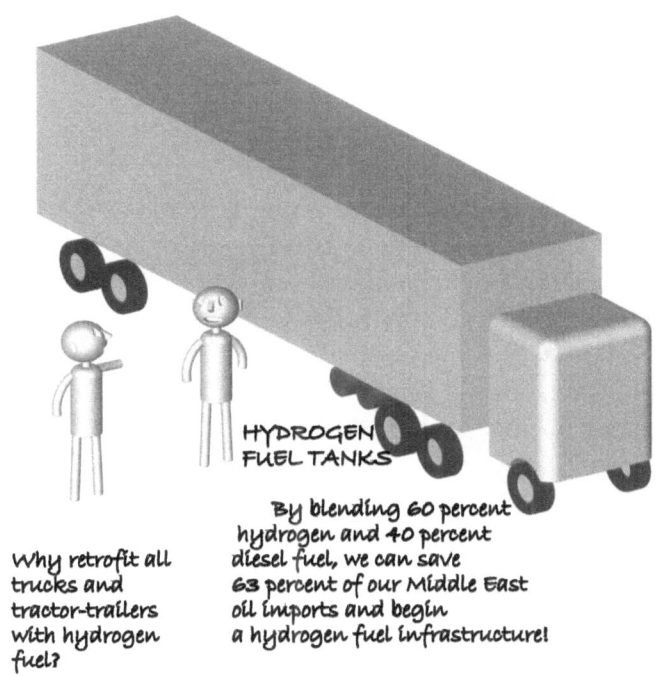

HYDROGEN FUEL TANKS

Why retrofit all trucks and tractor-trailers with hydrogen fuel?

By blending 60 percent hydrogen and 40 percent diesel fuel, we can save 63 percent of our Middle East oil imports and begin a hydrogen fuel infrastructure!

Imagine what would happen if there were a terrorist attack on a major Middle East oil reservoir. A 10 percent cut in world oil production could quickly send oil prices soaring. To solve this potential problem, the United

States is currently in the process of eliminating Middle East oil imports by emphasizing ethanol and trying to drill for more oil—like in the Arctic National Wildlife Reserve (ANWR) in Alaska and in the Gulf of Mexico. Although these efforts are a good idea, and may work after a long period of time, we may not have the luxury of time. Perhaps there is a way to get the job done faster.

The Plan

A faster way may be to implement a hydrogen fuel infrastructure for diesel-powered equipment like trucks, tractor-trailers, off-road equipment, boats, and trains, and do it in a three-year time frame. Why? Because 30 percent of our fuel usage is for goods and services provided by diesel-powered equipment. However, rather than converting existing diesel engines to run on hydrogen fuel only, we could make a compromise. And that compromise is to use a 60 percent hydrogen and 40 percent diesel fuel blend. This will allow existing trucks, tractor-trailers, and other diesel equipment to retain the option of having the same mileage range as they previously did with diesel only, if hydrogen is not available.

The 60:40 ratio is based upon research being done at the University of Tasmania. Their results show that electronically controlled blending of 60 percent hydrogen and 40 percent diesel fuel will result in 20 percent more power and eliminate nitrogen oxides. It is not clear how much additional mileage per gallon of fuel can be obtained; but with complete combustion with leaner mixtures during idle and at low-power operating conditions, a significant improvement should result.

Using the 60/40 mixture, we can replace 18 percent of our diesel fuel usage. And this amounts to 63 percent of our Middle East oil imports, or the equivalent of 3.3 Alaskan oil pipelines—let's shoot for 3.0 Alaskan oil pipelines since we probably won't be able to achieve a 100 percent conversion in the allotted three-year time frame. Replacing the remaining Middle East oil imports can be accomplished using ethanol and other methods.

At $120 per barrel of oil, the import revenue *saved* from three Alaskan pipelines (at one million barrels per day for each pipeline) would be $131 billion per year that can be used to produce and distribute the hydrogen fuel.

The proposed method for producing and distributing the hydrogen fuel begins with using domestically produced coal. And for now at least, the United States has plenty of coal. Here is how we can do it.

Let's begin with the fact that coal costs about $0.022 per pound, including transportation to a coal power plant. This contrasts with oil, which costs the oil companies about $0.43 per pound of oil that is undelivered and unrefined. However, the energy content of coal is about 25 percent less than oil. But the lower coal cost, even when adjusted for reduced energy content, is a very interesting fact. And it is this fact that will be used in our favor as we progress.

Next, we need to produce liquefied hydrogen at about eight hundred locations on or near existing coal-fired power plants. The process of producing the liquefied hydrogen will be by electrolysis and refrigeration—as described previously in our wind and water hydrogen scenario. We will use off-peak power from each coal plant to produce the electricity needed for the hydrogen fuel (I propose making hydrogen from water rather than directly from the coal, since it would help in getting the hydrogen infrastructure started and accomplish a second objective of retrofitting old coal plants with sequestering* technology). And since this is incremental power, the cost of generating electricity will be primarily a result of increased coal consumption—and not capital cost of the power plant. At 70 kilowatt-hours per kilogram of liquefied hydrogen fuel (equal to about one gallon of diesel

---

* *Sequestering* is the process of capturing carbon dioxide and piping it to underground caves for storage. I believe that this needs to be done because the public has been conditioned to believe in man-made global climate change.

fuel) and a power plant efficiency of 40 percent, we will need about 970 billion additional pounds of coal per year. At $0.022 per pound of coal, this equates to about $22 billion.

Figure 3.6.1. Liquefied Hydrogen Tanker Truck

Now here is where it gets interesting. Since the oil companies saved $131 billion in oil cost and spent $22 billion for our coal-to-hydrogen scenario, we have a net $109 billion remaining per year to retrofit each coal plant with sequestering equipment, purchase and install hydrogen production and delivery equipment (like the hydrogen tanker truck shown in figure 3.6.1), and retrofit diesel vehicles. At an estimated amortized cost of $21 billion per year to make this possible, we would have $88 billion leftover.

The Benefits

Perhaps the leftover $88 billion is savings that the oil companies can pass on to the public in terms of reduced fuel costs. And in case you were wondering who else benefits from implementing this plan, here is a list:

Coal companies double their yearly sales.

Freight train companies double their coal transportation sales and use the 60 percent hydrogen fuel blend.

Hydrogen fuel tank manufacturers sell 15 million hydrogen fuel tanks.

Hydrogen electrolysis equipment manufacturers sell twenty thousand electrolysis machines and drive production costs down.

Hydrogen liquefaction manufacturers sell seven thousand liquefaction machines and drive production costs down.

Electric power companies profit by additional electric power sales.

Electric power companies establish carbon dioxide sequestering systems at no cost to them or the public.

Oil companies profit from the sale of hydrogen fuel.

Trucking companies, individual truck owners, off-road equipment operators, and train companies realize increased engine power, less pollution, increased mileage, and hydrogen-use tax credits.

The United States and its citizens benefit from reduced Middle East oil imports by 63 percent in three years, reduced fuel prices, and tax revenue, resulting from the taxes paid by the above companies.

In case you did not follow what has been presented so far, you might want to go back and read it again. Just remember that we are using low-cost coal at *existing* coal power plants to replace high-cost imported oil. When we add up the numbers, we have some money left over. In addition, we can retrofit all existing coal plants with sequestering systems and have reduced diesel fuel emissions. And last, but not least, we will have the beginnings of a full-scale hydrogen fuel infrastructure!

And, consider this, the U. S. government could supplement the diesel truck conversion to hydrogen by retrofitting automobiles as described in the previous subchapter. Assuming that one hundred million automobiles were retrofit for three thousand dollars each, *the U. S. could save about 3 million barrels of imported oil per day or about $110 billion per year.*

If the U. S. government and oil companies would pay for this retrofit it would cost $300 billion plus the hydrogen production equipment and distribution costs. If the total cost was $400 billion, the amortized cost per year would be about $40 billion for ten years. This is a net savings of $80 billion per year to pay for the production of hydrogen. In my opinion hydrogen production should be the conversion of natural gas into liquefied hydrogen. *And, remember hydrogen will increase gas mileage by about 25 percent, versus essentially no increased mileage for using natural gas directly. This increased mileage should offset the increased processing and distribution costs. By the way, all of this money would be spent in the U. S. which would generate new tax revenue and new jobs.*

The Case for Trucks and Tractor-Trailers

Now let's discuss one aspect of the retrofit plan, that of converting trucks and tractor-trailers to hydrogen fuel. Retrofitting of off-road equipment, ships, and trains can be the subject of more discussion at another time.

Currently there are 15.5 million trucks and 1.9 million tractor-trailers operating in the United States. There are 360,000 trucking companies, and 82 percent of these companies operate six or fewer trucks or tractor-trailers. Therefore, we will concentrate our evaluation on the 1.9 million tractor-trailers and 82 percent of the trucking companies. For individually owned trucks and very small trucking companies, we can provide a special incentive for them to join the hydrogen club.

Now the question is, how are we going to convert these trucks and tractor-trailers to hydrogen fuel, and how are we going to make the hydrogen fuel available to them? Before we set up the hydrogen fuel infrastructure, we need to first understand the difference between a truck and a tractor-trailer. Trucks are generally used for short-haul deliveries and return to a depot each night for loading and refueling. Tractor-trailers, on the other hand, are generally used for long-haul deliveries and get refueled at truck stops along major highways. Given this general mode of operation, we can direct our hydrogen refueling locations to truck depots and major highway truck stops.

Before we go on, we need to discuss an important feature of long-distance tractor-trailer deliveries. Generally speaking, a tractor-trailer has enough fuel to travel about 1,000 miles before refueling (two large 84-gallon diesel fuel tanks located on each side of the tractor). If you remember our earlier discussion about liquefied hydrogen occupying four times as much volume as gasoline or diesel fuel, it would then follow that we would need to have a four-times-larger hydrogen fuel tank to equal the equivalent amount of diesel fuel. But since we are using only 60 percent hydrogen fuel, the tank volume would be reduced by 40 percent. We could reduce the tank size even more if we took the improved fuel efficiency into account.

Remember, in an earlier discussion, we stated that the Linde liquefied hydrogen tank, constructed from aluminum, weighed 220 pounds when filled with 22 pounds of liquefied hydrogen. Well, to provide the tractor-trailer with 1,000 miles of hydrogen-blended fuel, we would need two liquefied hydrogen tanks that are capable of containing 110 pounds of liquefied hydrogen. However, because of the insulated surface area to volume ratio for this increased volume of fuel, a full liquefied hydrogen tank would weigh about 440 pounds. This is five times as much hydrogen fuel for double the weight of the smaller tank. A tank of this capacity would be approximately 2.5 feet in diameter by about 7.5 feet long.

The two tanks would be attached under the trailer, thereby requiring a fuel tank connection when the tractor is mated with the trailer. Keep in mind that the tractor already has diesel fuel tanks that can power the tractor when it is separated from the trailer. The hydrogen mixture is introduced to the tractor engine with the flip of a switch. For large truck applications, the 440-pound hydrogen tank would also be mounted under the truck body. But since a truck does not require a manual attachment, the crossover is less of an issue.

Once the final configurations are resolved, the result will be a hydrogen fuel retrofit that does not require removal, or modification, of the existing diesel fuel tanks. This is important because each vehicle would be capable of running on diesel fuel only. By making the retrofits in this manner, the process could begin prior to having the hydrogen fuel infrastructure in place. This, then, could dramatically speed up the process of weaning the United States off its dependence on Middle East oil—in three years—and accelerate a hydrogen fuel infrastructure for all vehicles in the United States. In addition, because of the mini hydrogen fuel infrastructure, truck manufactures will begin making new trucks that use only hydrogen fuel; and electric power companies will build new coal power plants with a more efficient mode of power generation that includes carbon dioxide sequestering (i.e., integrated gasification combined cycle (IGCC) systems).

Hydrogen from Coal

Note: The Department of Energy (DOE) currently has a hydrogen-from-coal development program that has been in progress for more than three years. It may make sense to evaluate their progress relative to a continued near-term effort to implement the hydrogen infrastructure. As shown in subchapter 2.5, coal may run out during this century if a wind and water hydrogen infrastructure is not implemented. However, if clean burning techniques and carbon dioxide sequestering proves successful, deriving hydrogen directly from coal may be a better way to go for the short term. Perhaps the heat required to convert coal to hydrogen can be provided by the above-mentioned coal plant boilers.

3.7 The China Syndrome

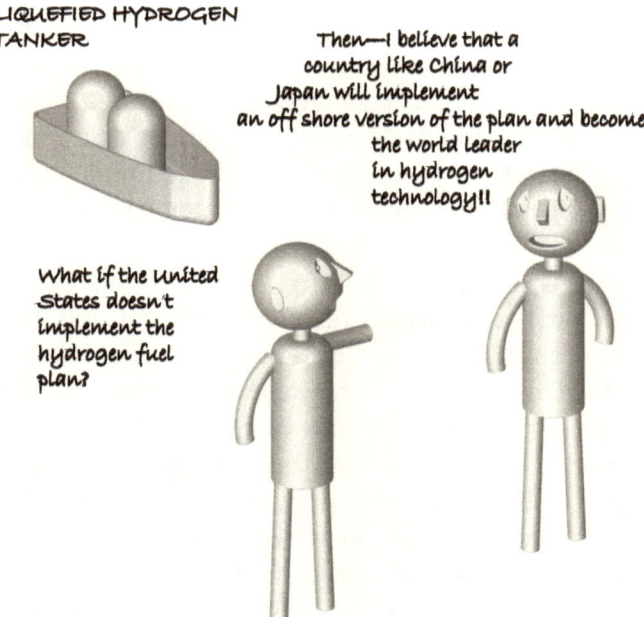

A movie was made in the 1970s called *The China Syndrome*, which was one of the contributing factors to abandoning "fission" nuclear power in the United States and in other parts of the world. Its premise was that a "runaway" chain reaction in a nuclear reactor would burn a hole through the earth and come out on the other side—in China. Although the title of this chapter is "The China Syndrome," its meaning is quite different. In this case, it is meant to alert people to the fact that if the United States chooses not to follow the recommendations outlined in this book, it would not be surprising to see these recommendations followed by either China or Japan. The result would be the same—a worldwide hydrogen fuel economy—but China and/or Japan would have a head start and a leadership position in the world.

Fortunately, the United States has vast underpopulated land area with high sustainable winds. But this is not the case in other parts of the world. This fact results in a new and different requirement for most other countries if they chose to create their own hydrogen fuel infrastructures.

The Ocean System

Fortunately, for most of these countries, like China and Japan, the oceans provide an answer. In the equatorial parts of the world, the ocean winds are not very strong, but the ocean thermal gradients (the temperature difference between the surface temperature and 2,000, or more, feet deep can be high and can be used to generate electricity and freshwater). Where the ocean thermal gradients are not very high, the winds are generally very strong.

In each case, an optimization of ocean waves, currents/thermal gradients, and strong winds could be employed. Ocean-based energy platforms would then be used to produce liquefied hydrogen and deliver the hydrogen to nearby ports via fast-moving catamarans, like the one shown in figure 3.7.1 below, which is being developed by Kawasaki Heavy Industries in Japan. An alternative would be to deliver electricity by means of superconducting underwater cable to onshore processing facilities.

Figure 3.7.1. Advanced Catamaran Hydrogen Transport Ship by Kawasaki

There are many possibilities that may be employed to capture the offshore wind, ocean movement, and/or thermal gradients. One possibility might be to use a wind turbine platform, like the one shown in figure 3.7.2. Using 600-feet-tall posts and 200-foot-long blades, this Norsk Hydro floating, and anchored, wind power platform is expected to be deployed in the North Sea in waters ranging from 200 feet to 2,200 feet deep.

Figure 3.7.2. Floating Wind Power by Norsk Hydro

A second idea would be to employ ocean thermal energy conversion (OTEC) systems like the one shown in figure 3.7.3. The OTEC system generates electricity utilizing the temperature difference between the warm surface-water temperature and the cold deep-water temperature. Between twenty degrees north or south latitude, the OTEC machine will produce continuous and reliable energy twenty-four hours per day all year long. Although its efficiencies are not very high—ranging from 2 percent to 8 percent depending upon the temperature difference—two side benefits are to generate freshwater, from seawater, during an evaporation phase and to bring deep-water nutrients to the surface that allows sea life to thrive—thus promoting improved fishing.

Figure 3.7.3. Ocean Thermal Energy Conversion System

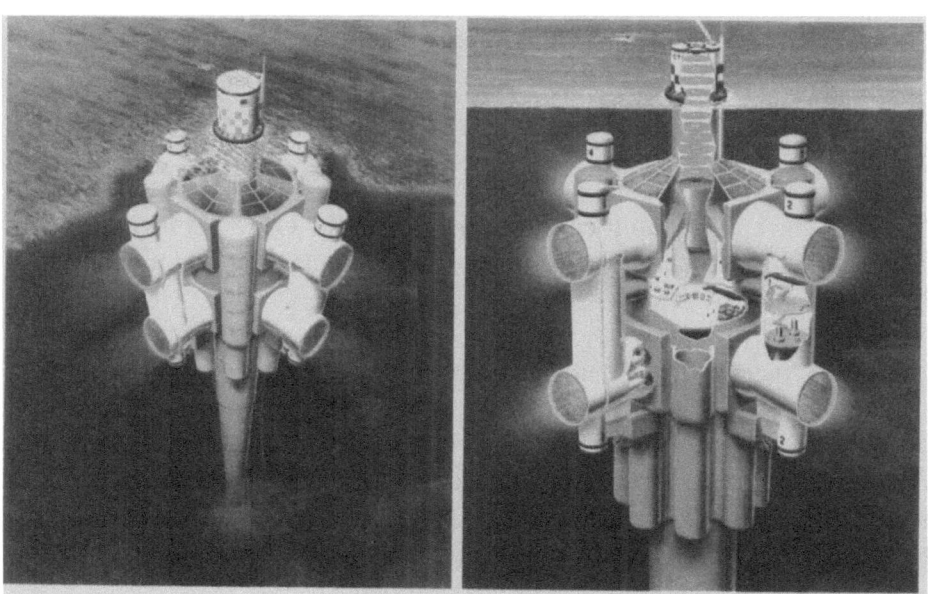

A third choice might be to have a hybrid system that combines an OTEC base with the previously mentioned wind turbine platform. Included in this hybrid may be a third system that uses wave or ocean current generators like those described in subchapter 2.4—more on this piggyback possibility later.

A Plan for Japan

I am not going to get into the design details, in this book, or detailed cost estimates for each offshore system. The purpose here is to make the reader aware of these options and of how other countries around the world will be able to make their own hydrogen fuel infrastructures.

I will, however, take a closer look at how a country, like Japan, might lead the world into the age of hydrogen fuel before the United States knows what is happening. This happened before when Edwards Deming—a U.S. citizen—proposed statistical process control and just-in-time manufacturing to the Japanese automobile manufacturers after being rejected by the U.S. automobile manufacturers. And you know what happened to our automotive industry. If you don't know, I recommend that you check it out on the Internet.

OK, let's look at the resources available to Japan:

A region of the North Pacific Ocean, close to Japan, has very high sustainable winds.

The best Pacific Ocean location in the world for OTEC is located about 1,000 miles southeast of the southern end of Japan, and this location also has moderately high winds.

Assuming that electricity could be transported underwater from the offshore generators, it would probably be the most cost-effective method of hydrogen processing. Therefore, because of its close proximity to Japan, the North Pacific may be the most prudent choice. In fact, the winds are so high and sustainable in the ocean adjacent to Japan's east coast, it may be possible to achieve 50 percent efficiency from low-altitude "breakthrough" rotor generators *to be described in the next chapter*.

Therefore, an estimate of the average electrical energy needed to turn Japan into a hydrogen fuel economy is about 352,000 megawatts. At 50 percent efficiency, the amount of installed rated wind turbine capacity would be 704,000 megawatts. This results in 35,000 20-megawatt generators. At 160 *actual* megawatts per square mile, we would need about 2,190 square miles of ocean area.

If we arrange twenty rectangle clusters, which are 15 miles wide and 15 miles deep, we would have 1,800 generators per cluster. Each cluster could have a one-mile-wide central shipping lane, maintenance access lanes, and probably four centrally located electrical processing and transmission facilities. Processed electricity would be sent to the mainland via superconducting cables for inland processing into liquefied hydrogen.

If the required twenty clusters were spread along the 912-mile coast of Japan, there would be a spacing of about 32 miles between clusters, which could serve as open-water shipping lanes.

Assuming that the price of the offshore 20-megawatt electric generators will be 30 percent higher than the land-based version, I have calculated the delivered liquefied hydrogen fuel cost to be about the same as the land-based version.

By increasing the energy density or by providing additional generator clusters, Japan could export liquefied hydrogen to other countries by using the previously mentioned catamaran tankers. In fact, for export purposes, Japan may elect to build additional generator clusters with OTEC bases in the South Pacific Ocean. By doing this, additional "buffer" and security backup would be provided from a distinctly different energy source location.

I don't think any more needs to be said.

Chapter 4
Things That Never Were

Some people see things the way they are and say why; I dream of things that never were, and say why not.

—George Bernard Shaw (later made famous by John F. Kennedy)

I was reluctant to include this chapter because the new ideas that will be presented are very different from anything that you may have read or heard about before. And since I don't have the luxury of having a research team available to me to verify the feasibility of various aspects of my proposals, I recommend that you regard them only as "food for thought" and let it go at that. As Yogi might have warned, "Don't make predictions—especially about the future."

4.1 Breakthrough

Yes—energy costs are about the same as conventional power plants and they require only three times the deployment space

Can that new wind generator compete with conventional power plants?

When Albert Einstein was asked how he was able to develop his revolutionary ideas, he responded by saying that he was not constrained by axioms. In other words, his revolutionary discoveries required him to think outside of the box. While I am no Einstein, I do try to see beyond conventional wisdom. Some readers may dispute what they are about to read (or have already read) because of things they have been taught to believe. Or in some cases, they have a vested interest in the status quo. Well, be that as it may, here we go with *"things that never were."*

Wind Power

Have you ever wondered why we don't have more wind- and solar-powered electric generation? To the average person, it is probably a mystery, since the wind and sun energy are free, and there is plenty of it to power the world. Even though the wind is not always blowing, and the sun is not always shining, it would still be sensible to make hydrogen and/or electricity in this way if it were cheaper to produce than conventional electric power generation. Sadly this is not the case; even though wind- and solar-power generation costs have been reduced by more than tenfold over the last thirty years, they are still more than two to three times as expensive to generate electricity as conventional fossil-fired or nuclear power plants.

Well, you are probably saying to yourself that if we wait just a little while longer, the cost will eventually become competitive. Perhaps you are right, but from my research and logic, the path that we are currently on, particularly with regard to wind power, will not get us to where we need to be. There needs to be new "out of the box" thinking. And it is not just cost that needs to be addressed; it is the whole wind power generation approach that needs to be revamped. That is what this subchapter is about.

Because of the inertia posed by the investments and subsidies that have been made in the propeller-and-pedestal technology, it will take a significant demonstration of an alternate technology, and proof of lower costs, to create a need to make the transition. And unless a government or wealthy investor puts up the money to make this demonstration, and provide this proof, it won't happen. *As a result, the propeller-and-pedestal investments will continue to become more entrenched as government subsidies propagate a bad idea.*

To begin, let's examine the largest propeller-and-pedestal wind turbine currently being built. It has a rated capacity of 5 megawatts. Its blades span 412 feet in diameter, and it is currently priced at more than $1,000 per rated kilowatt. Studies, by the CATO Institute, of smaller capacity propeller-and-pedestal installations reveal that only about 23 percent of their rated capacity is actually obtained. This is mostly because of a low pedestal height (as previously mentioned—a doubling of height generally results in about a 34 percent increase in wind power), not locating in a class 4 to 6 wind zone, and an inability to generate power at wind speeds of less than 10 miles per hour. *Therefore, the 5-megawatt-rated-capacity generator may actually yield about 1.15 megawatts, or about $4,350 per actual kilowatt produced. This results in more than double the cost of electricity produced by conventional fossil-fired or nuclear power plants, and the land (or ocean) space required for deployment is about twenty times more.*

To check on these CATO findings, I recently received a passout from the U.S. Department of the Interior regarding about 1,100 wind turbines located in the state of Minnesota, and the results were surprisingly different from what I had read in the CATO report. For instance, the wind turbines are rated at 1.65 megawatts and produce about 2 million kilowatt-hours per year. However, when you divide the 2 million kilowatt-hours by the number of hours per year, you get an average of 228.3 kilowatts per hour. This is only 13.83 percent of the rated capacity—a much lower percentage than that stated in the CATO report. To further exacerbate the situation, each of the wind turbines had an installed price tag of $2.5 million, which meant that cost per average kilowatt produced is $10,965. As a further bit of information, the pedestal bases are 16 feet in diameter and run 30 feet into the ground. And the total weight of each turbine is 2.9 million pounds.

Since most current fossil or nuclear power plants are rated at one thousand megawatts or more, these 1,100 wind turbines constitute the equivalent of about 25 percent of one conventional power plant! For now, at least, I will go with the CATO report numbers, which are still not very favorable.

I selected the Magenn floating air rotor because it had the potential of reaching 50 percent of its rated capacity and occupied about half of the deployment space. In addition, it would operate at much lower wind speeds and produce much less noise and electronic interference; it had a much less difficult field installation and minimal avian mortality (bird kill). Because of its lighter-than-air feature, it also had the potential of being much less costly when mass-produced. On the negative side, I was unsure about the ground-based mooring requirement—in the event of severe weather.

The Rotor Pedestal Design

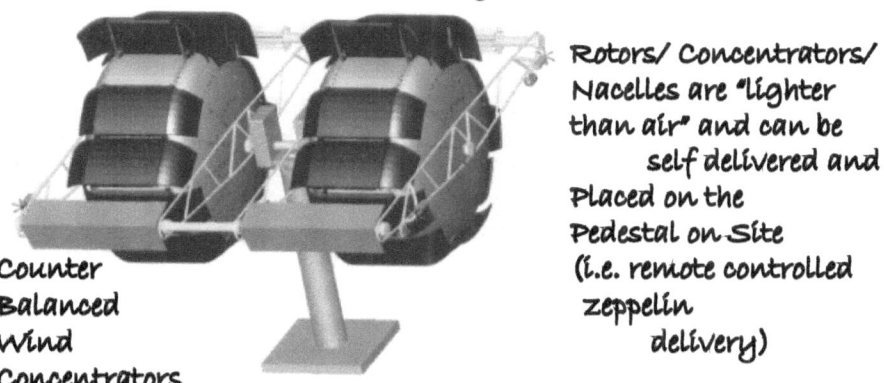

Two 10 Megawatt Air Rotors, Concentrators, and Nacelles on a Pivoting Pedestal

Rotors/ Concentrators/ Nacelles are "lighter than air" and can be self delivered and placed on the Pedestal on Site (i.e. remote controlled zeppelin delivery)

Counter Balanced Wind Concentrators

Propellers and Winched Cables Assist "Docking" on the Pedestal

Figure 4.1.1. 20-Megawatt Wind-Activated Electric Generator Using Conventional-Style Pedestal and Nacelles-430-Foot Closed Rotor Diameter

From observation, it is clear to me that the 5-megawatt wind turbine has probably reached its size limit, and from discussions with Magenn, their projected size limit is probably 10 megawatts. However, to make wind power really viable, the installed price, output capacity, and especially the deployment space needs to be dramatically improved. In this regard, I have tried to combine the best features of both the propeller-and-pedestal design with the best features of the Magenn air rotor. And the result is shown in figure 4.1.1.

As you can see, this design can produce 20 megawatts. I have estimated the efficiency to be about 40 percent, which results in an average yearly output of 8 megawatts. And because of its ability to be more closely spaced than propeller-and-pedestal wind turbines, it can produce as much as 128 actual megawatts in a one-square-mile space—probably the most important consideration. In severe weather, the blades will close to envelop the core zeppelin. Also, since the rotor generators are lighter than air, they can be brought to the deployment site by means of remote-controlled lighter-than-air self-delivery. In addition, a future version of this design might have four rotors rather than two. As a result, 144 actual megawatts can be installed in a one-square-mile space. This double-wide design would, of course, save money, but would need to accommodate increased stresses when the core zeppelins are not filled with hydrogen and are subject to hurricane force winds.

Now you may be asking, how did I determine 20-megawatt rating? Well, first of all, you need to see how the folding blades and concentrator work. This is illustrated in figure 4.1.2:

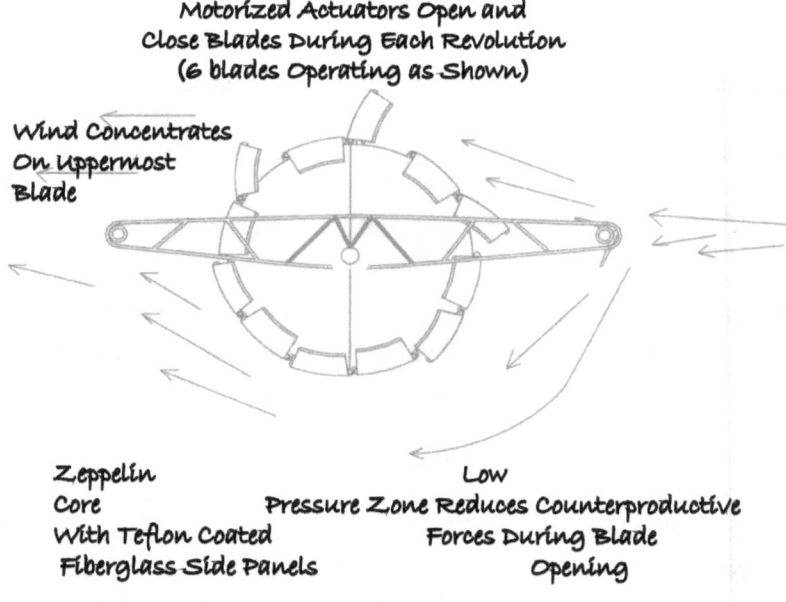

Figure 4.1.2. Operational Features of the Proposed 20-Megawatt Wind-Activated Electric Generator Concept—How It Works

The idea here is to minimize the counterproductive wind forces that result from blades that are traveling against the wind while making their circular path. Generally speaking, the propeller-and-pedestal design does not have to deal with counterproductive forces, but when these forces are eliminated from the rotor design, the advantage goes to the rotor. Why? Because if the topmost blade generates the power, and there are minimal counterproductive forces to oppose it, the power output is much greater. This is because the wind acting on the blade surface is transmitted to the shaft of the rotor by means of a torque arm. So the wind forces captured by the blade are augmented by the torque arm feature.

Let's put it another way. If we were to double the propeller-and-pedestal blade swept area, the capacity would increase about four times. In contrast, if we were to double the rotor size, the blade area plus torque arm length would increase capacity by about eight times (as long as we eliminate or minimize the counterproductive forces). It is this feature that has—in my opinion—misled engineers in the past with regard to rotor-type generators. So as I see it, the propeller-and-pedestal wind turbine was selected because of simplicity, and the rotor design was rejected because of its inherent complexity and counterproductive forces. *What this means is that propeller-and-pedestal wind turbines don't scale up very well, and my newly proposed rotor design does not scale down very well.*

What appears to have happened is that as the propeller-and-pedestal wind turbines became larger and larger, they reached a limit. On the other hand, as the rotor design increases in size, its output capacity offsets its greater complexity and cost. This combined with the lighter-than-air feature and the minimization of counterproductive forces leads me to believe that I am on the right track. In fact, my estimated cost of mass-producing and installing this rotor design is in the $2,000 per kilowatt ballpark.

Note: My calculated electric power for the rotor design does not take into account the wind concentrator or a secondary rotor generator. But to be fair, the calculations are based upon static and not dynamic wind forces and do not account for power losses resulting from the motorized blade actuator system, the remaining counterproductive blade opening forces, or the nacelle generator gear or electricity losses.

Magenn Redesigned

While designing the 20-megawatt rotor generator, the thought of giving up the 50 percent or more operating efficiency of the Magenn lighter-than-air design was somewhat disturbing. So to satisfy my curiosity, I developed the concept that you see in figure 4.1.3 below:

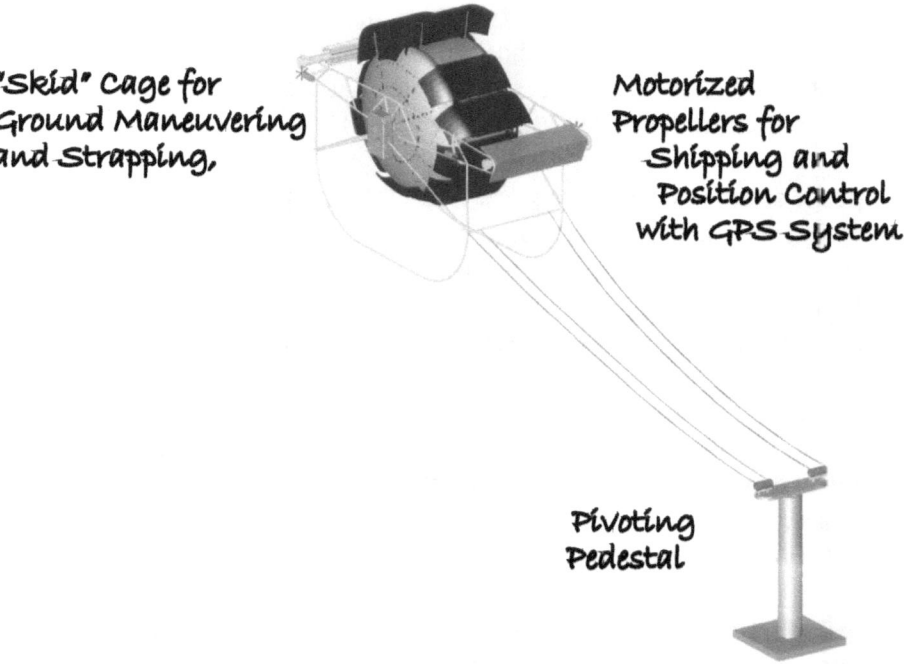

"Skid" Cage for Ground Maneuvering and Strapping,

Motorized Propellers for Shipping and Position Control with GPS System

Pivoting Pedestal

Figure 4.1.3. 10-Megawatt Air Rotor Operating at More Than 1,000 Feet Altitude-430-Foot-Diameter Closed Air Rotor

Operating at an altitude of more than 1,000 feet, this 10-megawatt version of the air rotor should achieve about 50 percent operating efficiency or 5 actually produced megawatts and 45 actually produced megawatts per square mile of deployment area. Since the previously discussed 5-megawatt propeller-and-pedestal wind turbine will produce only 1.25 actually produced megawatts and 18 actually produced megawatts per square mile, these advantages clearly favor this air rotor. And because one air rotor replaces about four of the 5-megawatt wind turbines, the price tag for the air rotor should be well under $2,000 per actual kilowatt produced—our pricing goal in subchapter 3.1.

As mentioned in subchapter 3.2, the Magenn air rotor could achieve a rating of 5 megawatts and operate at 50 percent efficiency. This resulted in 36 actually produced megawatts per square mile of deployment space and a mass-produced and installed price tag of $2,000 per actual kilowatt produced in a class 4 to 6 wind zone. However, as previously mentioned, I have uncertainties with regard to the proposed deployment space, the cost of ground-based equipment, and the $2,000 price tag. To overcome these uncertainties, I believe that my proposed 10-megawatt air rotor—above—could easily meet these requirements. *Note, however, that my 5-megawatt Magenn air rotor assumptions and my proposed 10-megawatt redesign have not been approved by Magenn.*

Design Calculations

So you don't believe that we can achieve the capacities or costs that I am projecting! Well, I have some doubts myself since I was surprised by the results of my static analysis calculations and the results of my mass production cost estimate. And after all, I don't have access to a design research team to validate my claims.

So before making a prototype, extensive dynamic (time-stepped) fluid mechanics analysis needs to be performed to optimize the design—a relatively low-cost exercise. It is my belief that the wind concentrator will need to be maneuvered for optimal performance during varying wind conditions and that additional parts—such as an updraft spoiler or focused ends—may need to be added. In addition, the drag coefficient may be improved through blade shape optimization and positioning. Although these features add to the complexity and cost, remember that we will be replacing about six 5-megawatt wind turbines with one 20-megawatt rotor.

Pricing

My pricing estimates are based upon having a very high volume. As discussed in the hydrogen fuel economy, described in subchapter 3.1, we will need about 170,000 20-megawatt generators in twelve years. This is equal to about forty-one generators per day that will, in my opinion, probably require forty-one factories producing one generator per day working twenty-four hours per day for seven days per week. As a result, for a $5-billion factory that is amortized for ten years, at 6 percent, there is plenty of money available to maximize robotic production without contributing a great deal of cost to each generator produced.

For example, if each factory were to cost $5 billion (a very high and conservative estimate), the ten-year amortized cost at 6 percent would result in a 16 percent increase in the cost of each generator. In addition, to keep the cost of construction materials low, I propose that we self-manufacture, from raw materials, all of the key material types—like aluminum, structural plastics, Kevlar, titanium, Teflon-coated fiberglass, and construction adhesives. And since the design is relatively simple and unsophisticated—but big—it can be top-down assembled with statistically controlled repetitive processes and robotic equipment. The only subassemblies that may require some degree of difficulty are the large 10-megawatt nacelle generators and the motorized system that opens and closes the "air scoop" blades. For a more complete description of the basis for how the proposed manufacturing costs were derived, please refer to "*Advanced* Mass Production" in the appendix.

Building It Big

Before leaving the subject of manufacturing, I am sure there are some readers who are rolling their eyes and saying, "Yeah sure." This thing is so big that it would be impossible to build or mass-produce—especially one generator per day. Well, history is filled with people who said it couldn't be done. If we had listened to these people, we wouldn't have gone to the moon, built jumbo jet airliners, or the Golden Gate Bridge. I am not going to describe in this book how manufacturing this behemoth could be done, but I have what I think is a workable solution. To give the reader some perspective of what we will need to do, I will mention a few interesting facts:

> The Golden Gate Bridge's towers are 770 feet tall, and the middle span is 4,200 feet long.
>
> The NASA space shuttle crawler is 131 feet long by 113 feet wide and is capable of transporting 12 million pounds.
>
> The University of Phoenix Stadium has a 403-foot-long-by-234-foot-wide 18.9-million-pound moveable field.

The University of Phoenix Stadium has a retractable one-sixteenth-inch-thick Teflon-coated fiberglass roof that is supported by two 700-foot-long-by-87-foot-high (at its midpoint) trusses. The top of the roof is 206 feet above grade level, and each roof panel weighs 1.1 million pounds.

The USS *Ronald Reagan* aircraft carrier is 1,092 feet long by 34 feet wide and weighs 190 million pounds when loaded.

The United States learned how to mass-produce 2,751 Liberty ships during WWII at a cost of under $2 million each (about $20 million in today's dollars). These ships were 441 feet long and 56 feet wide. Each ship was comprised of 250,000 parts that were prefabricated throughout the country. The average ship took seventy days to complete with the last ship—the SS *Robert E. Perry*—being built in four and one half days.

Also, as you may have guessed, a zeppelin transportation method could be used to carry a damaged generator back to the factory for repair and replace it with a new one. This minimizes downtime and centralizes the equipment and expertise needed to make the repairs. I have also developed some solutions for on-site repair that I will make available when, and if, this concept is developed.

To put the above discussion further into perspective, let's assume that the forty-one wind generator factories had a price tag of $205 billion. This is a lot of money, but when taken in the context of replacing 7.67 billion barrels of oil per year, at $100 per barrel ($767 billion per year), with hydrogen fuel, it doesn't take much thought to see the payback potential.

Consider This

At this point, it may be worth mentioning that the proposed 20-megawatt generator, shown in figure 4.1.1, is for land-based deployment. However, with some modification, it can also be adapted to ocean-based deployment with some additional cost. While the land-based version pivots 360 degrees on the supporting pedestal, the ocean-based version can be stabilized with a ballast base and use motorized propellers to provide rotation and control deployment position. Conversely, it can continue to use a pivoting pedestal.

With this in mind, you may recall the offshore economic analysis described in subchapter 3.7, which resulted in a similar liquefied hydrogen price to that of the land-based system. Although the rotor generator price was higher, we saved money because of an improved operating efficiency and no hydrogen pipelines.

Applying this thought process to the United States, many of the highly populated coastal states may want to take this approach rather than pipe in hydrogen from a remote Midwest location.

The Piggyback System

When finalizing the design of the 20-megawatt air rotor cluster, we should consider piggyback possibilities, since most of the supporting equipment would be paid for by the primary system that stands on its own merits. One system that comes to mind is wave generators. These wave generators could link the rotor generator platforms in a way that acts like a giant floating mat. As discussed in subchapter 2.4, underwater ocean currents could provide additional power. In subchapter 3.7, ocean thermal gradients were mentioned, but their use would probably be limited to between twenty degrees north or south latitude and would require deployment of the cluster in deeper water—beyond the continental shelf.

Another more controversial idea would be to employ small nuclear reactors as part of the rotor generator's ballast system. While you may be shocked by this possibility, you may not realize that current navy nuclear reactors use water to cool and moderate the nuclear reaction. Because of this feature, the reactor cannot have a "meltdown" when deployed in the ocean. This means that the reactor will self-extinguish in the event of an accident or a terrorist attack. Furthermore, the reactor cannot be damaged by an earthquake. Also, the navy reactors use very highly enriched uranium, which allows a very long refueling interval. Thus, their refueling costs are low, and their nuclear waste material can be compactly stored for a long period of time.

Since these small modular nuclear reactors can be mass-produced, quickly deployed, and piggybacked on existing equipment, their cost to produce electric power should be relatively low. If each reactor produced 20 megawatts, it would triple the electric power output from each floating wind-generating platform and thus reduce the number of deployed systems by two-thirds. Certainly, this would bring the cost of producing hydrogen down significantly. But here again, we need to be concerned about the depletion of nuclear fuel and the need for a breeder reactor.

Most breeder reactor concepts use liquid sodium as their coolant—not water. As such, the replacement of water reactors with liquid sodium-cooled reactors would need to be evaluated and adapted to self-extinguishing in an ocean environment. Along this line, look at the aforementioned base pedestal that I proposed for the pedestal air rotor. It is a concrete-and-steel structure that is about 60 feet in diameter and 400 feet high, and it doesn't do much of anything except support the air rotors and allow them to rotate.

If this structure were to house a 20-megawatt breeder fission reactor, it would serve as the reactor containment structure, a steam turbine and generator platform, and an evaporative cooling system—a cooling tower.

So there you have it, *"a thing that never was,"*—a *breakthrough* in how electricity and hydrogen fuel can be economically produced from the wind—and perhaps nuclear energy.

Disclaimer

I was surprised by the large megawatt capacities that resulted from my simplified static analysis of the rotor design. And as previously mentioned, I do not have a research organization to analyze and test my assumptions. Therefore, it is possible that these large capacities will not be realized for the sizes that I have selected. On the other hand, the capacities could be much higher because some of my assumptions are very conservative. However, we very much need to reduce wind power generation cost with deployment area being of particular concern, and the large rotor concept has the potential to do this. With this in mind, I believe that I am on the right track, and more work needs to be done to prove or disprove my claims. After all, if I am right, or close to being right, the result would be a *breakthrough*.

4.2 Lighter Than Air

Look at that huge zeppelin!

Yeah—its self powered and can lift one million pounds of cargo.

As you might have surmised by now, I find blimps and zeppelins to be very intriguing. And regarding *"things that never were,"* let's take a closer look at the zeppelin proposed in the last subchapter. Maybe it has other uses beyond shipping 20-megawatt generators from their mass production factories to their deployment sites.

The word *blimp* generally means "a lighter-than-air, balloonlike structure made of a leak-proof fabric." Whereas, *zeppelins*, like the *Hindenburg*, have a ridged structure with fabric on the outside and helium—or hydrogen-filled balloons inside. A compromise between the two would have a semiridged frame as part of the structure.

The Goodyear and MetLife blimps do not have a ridged frame and maintain their shape and buoyancy by inflating and deflating air bags as the helium gas expands and contracts with altitude and temperature. At ground level, the air bags are filled. When in the air, the helium expands; because the air is lighter at higher altitudes, the air bags are deflated. For these blimps, using a nonridged frame makes a lot of sense because of their relatively small size and marginal lift capability. However, as the size gets bigger, the lift capability is dramatically increased. This is because a doubling in size results in an eightfold increase in the volume of helium or hydrogen gas. And an eightfold increase in volume is about an eightfold increase in lifting capability. Therefore, a ridged frame can be used for larger sizes without much loss in the amount of cargo that it can lift.

To give an example of how this works, let's compare the Goodyear GZ-19 blimp to the *Hindenburg*. The GZ-19 is 192 feet long, 59.5 feet high, and 50 feet wide. The *Hindenburg* was 804 feet long and 135 feet in diameter. If you calculate the volume of the GZ-19, it is about 300,000 cubic feet, while the volume of the *Hindenburg* is over 7 million cubic feet—an increase of over twenty-three times that of the GZ-19—enough so that a ridged frame is practical. And by having a ridged frame, it could lift a substantial amount of weight without damaging the balloon material. In the case of the *Hindenburg*, it was capable of lifting about 247,000 pounds in addition to its own weight. See figure 4.2.1.

Figure 4.2.1. The *Hindenburg*

So in case you were wondering how big the zeppelin was—that we described in the last subchapter—I will tell you that it is much bigger than the *Hindenburg*! Our zeppelin is 400 feet in diameter and 180 feet high. And utilizing the *Hindenburg*'s weight as a base for calculating the larger zeppelin's weight—on the basis of four increments of weight increase for eight increments of volume increase, the net lift capability of the larger zeppelin is 1.5 million pounds. My research shows that this is 0.5 million pounds more lift capability than the largest publicly known zeppelin that I could find on the Internet. Yet the proposed larger zeppelin is 65 percent shorter than the *Hindenburg* but 288 percent larger in diameter. As you can see, this size increase equates to more than six times the lifting capability of the *Hindenburg*. In addition, where high altitude is required to cross mountain ranges—let's say, 25,000 feet—the net lift would still be over 0.5 million pounds.

Let's pause here for a moment and make some things clear about why large zeppelins should be reconsidered. First of all, there are new technologies that were not available when the zeppelins were discarded. The second reason is that we need a low-cost alternative to current methods of transporting large objects—such as large wind-activated electric power generators.

So what are the new technologies? Well, how about global positioning systems and computer technology that permits unmanned operation, thus permitting the possible use of low-cost hydrogen, rather than high-cost and scarce helium, as the lighter-than-air gas. The next technology is photovoltaics. Because of the large top and side surface area of these behemoths, we could use photovoltaics to supplement the propulsion fuel, thus making their operational cost much less than current transportation systems. In addition to new materials such as Kevlar, titanium, carbon composites, and aluminum alloys, another technology is the current use of computerized design analysis methods that are capable of evaluating wind forces and structures in a way that was not dreamed about in the days of the *Hindenburg*. Then, while mass production was available when zeppelins were discarded, there was no need to mass-produce them in the numbers required for deploying the wind-activated electric generators. Additionally, robotic assembly and my proposed advanced mass production method— described in the appendix—were not in use at that time. Production cost would also be reduced by using the same factory and construction methods used to produce the wind-powered electric generators. Then by hardening the zeppelin surface with photovoltaics, it would not be necessary to have expensive ground-based enclosures. Instead, a low-cost strapping system could serve to brace the zeppelin during severe wind conditions. In fact, because of the cylindrical shape of the proposed zeppelin, it would not need to be rotated to face toward the oncoming wind. Deflation of the zeppelin, in very severe conditions, could also be a solution.

Because of using a photovoltaic surface, our zeppelin's cost is increased, and its lift capability is decreased—but it will still be capable of lifting one million pounds. The net result being a mass-produced, low-cost, high lift capacity device that is capable of relatively inexpensive air transportation. Where high speed is not a requirement the proposed zeppelin could be self-powered. In cases where higher speeds are required, we could, of course, provide additional fuel tanks and a higher horsepower propulsion system.

Can you think of some other uses for this zeppelin concept other than transporting the 20-megawatt generators? How about carrying cargo that would normally be transported by truck or tractor-trailer? Have you ever wondered why the premanufactured house has never caught on? It is probably because the majority of the cost is in transportation and site preparation (e.g., basements, streets, sewer lines, electrical lines, water lines, and landscaping). Well what if the basements, sewer lines, etc., were premanufactured? And what if we could do some of the site preparation with our zeppelin instead of dump trucks? Then we could deliver the house with basement, landscaped front yard and backyard, paved street, and utility hookups in one move!

What else? How about having the capability of dumping four Olympic-sized swimming pools of water on a forest fire or carrying the load equivalent of two superlarge mining dump trucks at a fraction of the cost.

Our zeppelins could also serve in the area of disaster relief. For instance, we could help evacuate cities like New Orleans before a hurricane and flood. Or how about providing water to drought-stricken areas or food and housing relief in case of an emergency.

How about premanufacturing concrete aqueducts to help increase the world's cropland? And in our hydrogen fuel scenario, we could help prepare the land and deliver premanufactured hydrogen pipelines.

Surely, the military could use a low-cost method of transporting troops and equipment. And by keeping a large number of zeppelins continuously operating at high altitude, supplemented by hydrogen-filled Kevlar weather balloons, we could replace some very expensive communications and surveillance satellites.

Finally, how about using the zeppelin to help with worldwide recycling of our nonrenewable resources? I will discuss more about this in another subchapter.

I am sure that you could think of other uses, and that is why the mass-produced, low-cost, large-scale, semi-self-powered zeppelin is a *"thing that never was"* but perhaps is a thing that should be.

As a final note, I want to clear up the subject of zeppelin reliability. Besides the *Hindenburg* disaster, there were two other notable zeppelin disasters that need to be discussed. They are the U.S. Navy's *Akron* and *Macon*. They were almost as large as the *Hindenburg* and had the additional advantage of being filled with helium rather than hydrogen. Briefly, these zeppelins were destroyed while operating in severe weather that damaged their structures and caused their helium-filled bladders to deflate.

Well, that was then, and this is now. Back then we did not have the engineering analysis tools and test equipment that are available today. So without going into a lot of technical detail, suffice it to say that if we can design a highly reliable jumbo jet aircraft, we can design a reliable zeppelin—even one that will operate or survive on the ground in severe weather. After all, the *Hindenburg* and its predecessor, the *Graf Zeppelin*, logged over one million miles during their time in service. And keep in mind that before we were able to go to the moon, we had a number of launchpad disasters before we figured out how to do it.

4.3 The Net Minus Effect

Consider the colossal beauty that Disney created from marginal land in the middle of Florida. Then envision a large portion of the vast desolate lands of the world being transformed into beautiful forests, lakes, hills, valleys, flowers, gardens, rivers, streams, waterfalls, and futuristic cities and towns. At the same time, imagine this vision blending in with the beauty of the desert and providing environmental habitats for protected species. Now keep that vision in mind as you read this subchapter about "*a thing that never was,*" the "net minus" effect.

Before we begin our utopian odyssey, I need to state that I am a skeptic of man-made global warming. There are too many other factors to consider before making the assertion that man-made carbon dioxide is the cause of increasing the world's temperature. For instance, variations in the sun's radiated temperature, changes in the earth's magnetic field, heating of the oceans from varying heat cycles within the earth's liquid core, and changes in the amount of heat that is reflected or absorbed by the earth. With that being said, although current man-made carbon dioxide currently constitutes less than 2 percent more than naturally occurring carbon dioxide, the man-made climate change theory probably should not be dismissed if the hydrogen solution could result in a worldwide economic and environmental improvement.

So far, my solution to the transportation fuel crisis has been to convert to hydrogen fuel as soon as possible. While this solution significantly reduces the amount of future carbon dioxide emissions from transportation vehicles, it doesn't solve all of the sources of man-made carbon dioxide emissions. Hopefully, these other emissions will eventually be eliminated with the introduction of cheap and abundant wind, solar, and breeder fission and/or fusion nuclear electric power. However, during the transition time period, the world will be spewing out additional amounts of carbon dioxide and pollutants into the atmosphere.

Generally speaking, it takes fifty to two hundred years for atmospheric carbon dioxide to go away though natural causes (absorption by the oceans, soil, vegetation, and the biosphere). Therefore, a prudent thing to do might be to introduce a *new* worldwide carbon dioxide absorption system. And since a natural carbon dioxide absorption system is a plant or a tree, it would seem that all we need to do is produce more plants and trees. But wait a minute, when plants and trees die, their decaying process gives back the carbon dioxide that they had absorbed. In other words, a "net zero" effect on carbon dioxide removal. However, one solution to this "net zero" equilibrium process is to sequester the carbon dioxide during the decaying process, thus making the plant or tree a "net minus" contributor to reducing carbon dioxide from the atmosphere. Note, however, that trees can last for many years, thus delaying their carbon dioxide release.

To give the reader a rough idea of what this "net minus" effect is, consider that about 5 million pounds of carbon dioxide would be removed from the atmosphere per square mile of vegetation per year. To put this into perspective, the world currently spews about 53 trillion pounds of man-made carbon dioxide into the atmosphere each year. However, only 57 percent—or 30 trillion pounds—is retained by the atmosphere (called the airborne fraction). To equal this rate, we would need 6 million square miles of vegetation. Putting this into perspective, the entire land area of the United States is 3.54 million square miles.

But wait, we really don't need to equal the spew rate, since over about fifty years we can reduce the spew rate close to zero with our hydrogen fuel and "fusion" nuclear plan. So depending on the amount of land used for carbon dioxide removal, at some point in time, over the next fifty years, we will have a crossover. With this in mind, let's assume 0.5 million square miles of worldwide land is available for new carbon dioxide-removing vegetation. This 0.5 million square miles would remove about 2.5 trillion pounds of carbon dioxide per year—or 8 percent of the current man-made carbon spew rate. This is not insignificant, since the current spew rate will go up at first and then drop to near zero in less than fifty years—if the hydrogen fuel and fusion/breeder fission nuclear scenario prevails.

As you may have guessed, I am not talking about just planting trees; instead, I am referring to planted vegetation that has a high yield of ethanol relative to the energy expended to produce it. As you may recall, I did not believe that the transportation fuel problem could be solved using ethanol. Although ethanol is a renewable energy resource, it requires too much valuable farmland that should be used to help feed the world's growing population. However, I did not discuss the possibility of using desolate land that is currently not suited for growing ethanol vegetation. *Growing ethanol vegetation on desolate nonfarmland would add to the world's existing carbon dioxide-reducing vegetation.*

Switchgrass has been mentioned as a high-yield cellulostic ethanol producer that can be grown on marginal land and doesn't require much water, fertilizer, or maintenance. So let's begin with switchgrass. As a starting point, we will look at a scenario just for the United States and then compare the result with the previously discussed hydrogen fuel scenario.

It might surprise you to know that the United States Government owns 415,000 square miles of desolate land in the lower forty-eight states. Much of this land is in the state of Nevada, but includes the states of Wyoming, Utah, Idaho, Montana, North Dakota, South Dakota, Texas, New Mexico, Arizona, and California. If we assume that 150,000 square miles of this land can be converted to ethanol-producing land, it would equal 40 percent of the current United States transportation fuel consumption rate. Now let's add 50,000 more square miles for the previously mentioned Disney stuff—forests, lakes, hills, valleys, gardens, rivers, streams, waterfalls, nature preserves, and futuristic cities and towns. Other desolate land may also become available as the desirability of our plan is revealed. For instance, 85,000 additional square miles of land is home to about five hundred Native American tribes.

Using 40 percent of the United States transportation fuel consumption rate as our goal, we can subtract 40 percent of the cost of our previously discussed hydrogen fuel from water scenario (subchapter 3.1) to help determine a potential overall expenditure limit. However, to keep our "net minus" goal for carbon dioxide absorption, we need to *reform the ethanol into hydrogen* and sequester the carbon dioxide resulting from the reforming process.

Before we discuss how we are going to transform the desolate land into ethanol-producing land, we need to make one further assumption. The assumption is that all energy used in processing the ethanol is produced by our 20-megawatt hydrogen-from-water wind generator systems. Everything from pumping water to reforming the ethanol into liquefied hydrogen would be included. For switchgrass, this electric energy is conservatively assumed to be equal to the energy produced to make hydrogen from water and needs to be available on a seasonally adjusted basis. Which means that most of the required energy is used during the seasons when cellulose pretreatment, distillation, reforming to hydrogen, and carbon dioxide sequestering processes are implemented. Because of this requirement, we will assume that the number of wind energy generators will be equal to the number used when producing all of the fuel as hydrogen from water. What this means is that of the 170,000 20-megawatt wind generators required for the hydrogen-from-water scenario, 40 percent, or about 68,000 generators, would be devoted to the ethanol-to-hydrogen scenario. As a result, there will not be a reduction in the energy-producing equipment costs. But there will be an overall increase in the amount of energy produced.

Without making this more complicated than it really is, just keep in mind that we used 78 kilowatt-hours of energy to produce one kilogram of liquefied hydrogen fuel from water. In our equal-energy ethanol scenario, we only need 36 kilowatt-hours of energy to produce one kilogram of hydrogen fuel (one kilogram of hydrogen fuel is equal to 36 kilowatt-hours of energy, or one gallon of gasoline). So by keeping the same number of wind energy generators, we can have enough power available to account for the seasonal energy peaks. But there is a bonus; we can now use the nonpeak excess energy to sell electric power to the National Electric Grid. This excess "green" electric power further reduces carbon dioxide and polluting emissions.

You may want to read the last two paragraphs again if it was not clear the first time through. But as you might surmise, the economics of this ethanol scenario might result in a net cost reduction. Let's find out.

For our economic scenario, we are going to obtain most of the land from the United States Government, at no cost, and then we are going to give it away to families who qualify for relocation. This is where it becomes very interesting. Let's suppose that we gave one-half of one square mile of land to each of 300,000 qualifying families. In addition, we gave each family $300,000 to get started. They would use this money to help purchase farm equipment and build a house. Each year thereafter, the family would have a gross income of about $180,000, if we assume $20 per equivalent hydrogen barrel of ethanol vegetation. Not too bad a deal, but the giveaway just cost us about 12 billion per year when amortized at 6 percent over ten years. And that's not all. We now need to add the cost of getting water, soil, and fertilizer to the desolate land locations, cellulose pretreatment facilities, distilleries, hydrogen reforming equipment, carbon dioxide sequestering facilities, pipeline transportation to central distributors, and infrastructure development. Because of the additional controversy that may develop from trying to estimate these costs, I am going to say that this entire cost is equal to, or better, than the cost of producing all hydrogen fuel from water.

On the positive side of the ledger, remember that we are getting a payback from the sale of excess electricity to the National Grid. We will also eliminate twenty of the proposed fifty hydrogen gas pipelines and replace them with ten single small-diameter ethanol pipelines. Without going into the cost trade-off of storing hydrogen versus storing ethanol, it goes without saying that ethanol would be easier, and less expensive, to transport and store than hydrogen.

On the negative side of the ledger, we are not just putting in infrastructure such as roads and sewerage systems, we are going to do the Disney thing. Trees, lakes, hills, valleys, flowers, gardens, rivers, streams, waterfalls, environmental habitats, and futuristic buildings will be part of the final design.

And now for the hard part, where do we get all of the water, and even soil and fertilizer, that will be required to make this utopian landscape possible? If you do the calculations, you will find that, even switchgrass needs lots of freshwater, soil, and fertilizer for 150,000 square miles. So here, we have choices to make.

Did you know that groundwater comprises about fifty times more freshwater than can be found on the earth's surface? In fact, the Ogallala groundwater aquifer located under portions of eight Midwest states is one of the world's largest underground reservoirs. It lies under 174,000 square miles of land located to the east of the Rocky Mountains, which is often called the "bread basket of the United States." The water depth ranges between 100 and 400 feet and is between 3 and 525 feet below the High Plains surface.

At first, you may be thinking that this water could be piped over mountainous terrain to the previously mentioned 150,000 square miles of desolate land area. It probably can, but there is a problem. The freshwater from the Ogallala aquifer is currently being consumed about ten times faster than it is being replenished. This is due to evaporation by the High Plains arid atmosphere, the impermeability of the top layer of soil (called caliche), and small pore spaces in the underground natural resupply network. This problem is not unique to the Ogallala; it generally exists for other aquifers throughout the world as well. While it is not clear how much longer the Ogallala will continue to provide its vital water, what is clear is the increasing cost of drilling and pumping deeper and deeper wells. And when the water supply cannot keep up, neither will the food supply. Have you ever heard about the semiarid High Plains crop failures in the 1930s due to cycles of draught, culminating in the disastrous Dust Bowl? This event preceded the development of cheap and efficient electric pumps to extract water from the aquifer.

Recognizing that streams and rivers eventually feed their freshwater into the saltwater ocean, there may be a way of diverting this water for our purposes. Through placement of flow-control dams, it may be possible to extract large quantities of water from a stream or river. Assuming that we were to extract this water with pumps and pipes, it would require an estimated seventy 12-foot-diameter pipes to supply enough water for 150,000 square miles of desolate land for ethanol production. A massive, and potentially very costly, project, but not impossible, especially when the pumping power is provided by our wind power generators.

Here we have a dilemma. On one hand, we are letting the freshwater from our rivers flow into the saltwater ocean—effectively throwing it away. While on the other hand, we are depleting our vital fresh groundwater supply. There must be engineering solutions to this dilemma that have been considered but have not been implemented. This is probably because of a belief that the aquifer problem can be controlled with more efficient farming and watering techniques now, with the real problem of aquifer replenishment to be dealt with in the future. Well, we have a fuel shortage and a potential global warming problem that can't wait much longer. So why not "bite the bullet" and deal with both of these problems now?

I am not willing to speculate, in this book, on how a cost-effective and environmentally friendly engineering solution can be implemented. But suffice it to say that creating high-mountain lakes in the Sierra Nevada and/or Rocky Mountains from existing rivers and streams may be a good starting point. However, during this process, innovative environmental protection solutions would need to be implemented. Think about it.

Now let's deal briefly with the soil and fertilizer issue. Do you remember the Rentech Company (mentioned in subchapter 2.2) that produced liquid hydrocarbons as a by-product of making fertilizer? So why not use our extensive coal deposits to make hydrogen and fertilizer. We would, of course, sequester the carbon dioxide during this process. Also, do you recall the large-scale zeppelins (from the previous chapter) that could be used to deliver soil and fertilizer?

So there you have it. A 40 percent ethanol-to-hydrogen scenario for the United States that *may* cost about the same as converting water into hydrogen. But this ethanol-to-hydrogen scenario has the added benefit of reducing atmospheric carbon dioxide, storing hydrogen fuel in the form of ethanol, providing homes and jobs for millions of people and maybe—just maybe—replenishing and saving our West, and Midwest, aquifer water supply. If this works, other countries may do a similar thing and provide an additional 350,000 square miles of ethanol-producing desolate land. If they do, significant amounts of hydrogen fuel would be produced.

4.4 Intended Consequences

Did you know that 35 percent of the world's population lives without electricity and 70 percent don't use gasoline?

Yeah—but what happens when when they do?

If you were to examine new government programs, you will probably find that, in many instances, some aspects of the programs have not been fully anticipated. As a result, unintended consequences cause the programs to either fail or become too expensive. This is especially true where the government provides payments that are intended to either help people in need or to stimulate something that it feels needs to happen—like subsidies for corn ethanol.

In this book, I have discussed how to fix the problem of running out of fossil fuels. However, I have not addressed the potential problem that a growing, more industrialized population presents. And that is an ever-increasing use of other nonrenewable natural resources. In this subchapter, I will try to make a "connection" between both problems and use the solution to the fossil fuel problem to help solve the other.

With the about 20 percent of the world's population currently consuming most of the world's nonrenewable natural resources, it is logical to assume that when, and if, the rest of the world catches up, there may not be enough to go around—especially when most of these resources are being thrown away.

Here are some previously mentioned facts that should give cause for concern:

> Half of the world's current 6.6 billion people live on less than the equivalent of one U.S. dollar per day.

> The 4.6 million square miles of the world's farmland has lost much of its productivity since 1945—an area about the size of India and China combined.

> Farmers are abandoning about twenty-seven thousand square miles of farmland each year because of lack of water and soil degradation.

> The world's fresh groundwater is being consumed about ten times faster than it is being replenished.

> The world's population will grow to about 9.2 billion people by the year 2050, the majority of which being in underdeveloped, nonindustrialized countries.

> The underdeveloped, nonindustrialized countries are rapidly becoming industrialized and are consuming more and more of the world's nonrenewable resources.

A pessimist may say that the world cannot sustain this population growth with its increased nonrenewable resource consumption. An optimist may say that we can sustain this growth with new technologies and innovations that have not yet been proposed. A pessimist might try to solve these problems by impoverishing the rich in an attempt to help the impoverished. An optimist may say that we will discover new technologies that will enrich the impoverished.

As you already know from reading this book, we can provide enough electric energy and transportation fuels for a much larger world population using my proposed wind-activated electric generators, fusion and/or breeder nuclear power, and hydrogen fuel. By using these energy sources, we can bring the issue of man's supposed contribution to global warming under control. And the further industrialization of the world's population will tend toward a stabilization of population growth.

So rather than denying the growing world population from participating in our modern industrial achievements and medical breakthroughs, we should welcome them for our benefit and for the benefit of the planet. But there is a need to restore farmlands and recycle almost all nonrenewable natural resources. Here is how the solutions used to solve our energy problem can help in providing solutions to solve our nonrenewable natural resource-depletion problem:

Restoring Farmlands

The subject of restoring farmlands will be discussed here briefly since I am, to say the least, not an expert. However, research shows that the problem can be solved by reducing erosion due to water runoff and by providing natural nutrients to the soil.

It has been demonstrated that using "contour bunding" to reduce water runoff, and digging holes to retain manure and water in the plant root zone, will successfully restore degraded farmlands. This proven process began with raising cattle and other farm animals to provide manure. The manure stimulated more plant growth that allowed more farm animals, and with more farm animals, there was more manure in a self-reinforcing cycle of land renewal. The process took about twenty years but resulted in substantial productivity gains and food security. In addition, when growing crops on the restored land, a process of crop rotation and drip watering can prevent a new process of soil degradation and preserve water.

The process of providing water to drought-stricken land will require some imagination. However, one method is to divert water from distant locations to the drought-stricken areas by several means—one of which is by using the previously discussed large-scale zeppelin to build aqueducts.

How the farmers survive during the twenty-year land renewal time period depends upon them having other sources of income and the availability of some imported food. And that is what I will discuss next.

Worldwide Recycling

At some point in time, worldwide recycling of nonrenewable natural resources may not be an option. However, nonrenewables can be classified in two forms. The first form being fossil fuels and the second being minerals. In the case of minerals, an eventual solution may be found in our solar system—like mining the moon and/or asteroids. While mining the solar system may be possible in the future, it may not come soon enough, and the economics of recycling what we already have on earth may still be necessary. Needless to say, if 9 to 11 billion people were to use nonrenewable resources, the way that the industrialized world uses them today, most of these resources will eventually end up in landfills.

In my opinion, worldwide recycling needs to begin with new product design. In today's world, engineers think more about making a new product that will sell more than what will happen to the product when it wears out. In fact, many times there is little concern for the repair mechanic that needs to service and repair the product.

With recent advancements in robotics, it would seem prudent to design new products with robotic assembly in mind. By doing this, the designer could also consider robotic disassembly. *In a perfect recycling world, products would be produced on an assembly line, with worn-out products being returned to the factory for processing on a disassembly line.*

Since the original product manufacturer knows more about the original product's material pedigree and assembly process, that manufacturer would be best able to perform the disassembly and reuse process. Also, since the original product manufacturer would know the material specifications for self-manufactured parts and subassemblies, they would also know the companies that supplied purchased subassemblies and parts. These subassembly and part manufacturers could then return *their* purchased subassemblies and parts to the originating companies to the point of raw materials processing.

You are probably saying that this is "pie in the sky" thinking and that this so-called perfect recycling world would be so impractical that it will never become a reality. Maybe so, but this is where I dream of *"things that never were"* and say, *"Why not."*

OK, let's examine the advantages to companies that do participate in this process. First of all, products that are assembled and disassembled by multipurpose robotics would tend to be less expensive—even when compared to low-cost manual labor. Second, if the manufacturer planned for disassembly of obsolete products with the potential for reusing the obsolete product's materials and subassemblies in their new products, they could further reduce costs. Third, the product manufacturers would use more noncorroding materials such as structural plastics, carbon composites, aluminum, stainless steel, titanium, copper, and brass, which would make their products lighter weight, longer lasting, and potentially higher quality. And last, but not least, the product manufacturers could restore (remanufacture) obsolete products for low-cost resale to emerging countries. In fact, one scenario may be to make a new product from old parts. Even though remanufacturing automotive parts is a big business today, what I am talking about is remanufacturing the entire automobile (or other product). People in emerging countries are not as enamored by up-to-date products as they are in being able to afford a product that they didn't have before.

Now you're saying, yeah, but how are we going to rescue the obsolete products and then return them back to each and every manufacturer without incurring a very high cost and complexity? Well, have you ever heard of Federal Express or UPS? These companies generally deliver packages to central locations for sorting, and then they ship the sorted packages to their destinations. Some packages that are destined to go to a nearby location are sent to distant central locations first. Believe it or not, this turns out to be the least costly way to do it.

But doesn't all of this transportation of products cost a lot of money and use a lot of fuel? Yes, if done by conventional means, like air cargo planes and trucks—especially when we plan to pick up and deliver some very heavy products and deliver them to distant locations. OK, hold on to your seats; remember our previous discussion about the large scale, semi-self-powered, slow-moving zeppelin? Well, that is what I propose that we will use in this scenario.

So what does all this have to do with the billions of poverty-stricken people around the world who are trying to survive on degraded farmland? Well, during the twenty or so years needed for farmland restoration, we could employ the services of the farmers of this land to process manufactured products for shipment back to original manufacturers—they would be the Federal Express central processing and sorting locations. However, since current product manufacturers are not set up for returned products, I propose using these locations to manually disassemble products and sort the product parts by material groups. The material groups would then be returned to raw material manufacturers for reprocessing. As time goes by, the original product manufacturers can become part of the program in an attempt to reduce their product cost. After all, if they don't, they will continue to pay ever-increasing market prices for materials that they had previously purchased. Get the point?

You have probably heard the expression "Give a person a fish and the person will have enough food for a day, but give that same person a fishing rod and the person will have enough food for life."

In my plan, we are going to invest in recycling "fishing rods" with the intent of stimulating an economy that will pay itself back many times over.

Let's begin with fifteen thousand distributed locations around the world that will act as central processing and sorting locations for obsolete manufactured goods. In our case, however, we will begin with the reprocessing of automobiles with aluminum block engines (about 45 million automobiles become obsolete each year). Although about 80 to 90 percent of aluminum engine blocks are currently being recycled worldwide, we will underprice the current recyclers. This may seem harsh to those who do this recycling now, but keep in mind that we are trying to achieve a much larger objective. Perhaps a onetime buyout of each company affected may be a reasonable approach.

At $0.50 per pound of aluminum and about 300 pounds of aluminum per car, we could earn $150 per car for the aluminum block alone, and additional income would come from other parts that are removed from the car. For the 45 million cars that become obsolete each year, we could process 60 cars per week at each of 15,000 locations. Now assume 60 workers per location, who can clean and disassemble about 60 cars per week and are each paid considerably more than $150 per week because of recovered parts in addition to the aluminum blocks. As you can see, the aluminum engine block alone could support this process—assuming there were no other operational expenses! And of course, one significant operating expense would be electric power and fuel.

Therefore, we are going to generate all of the required electric power and fuel using the mass-produced, low-cost photovoltaics and hydrogen electrolysis equipment developed while implementing the hydrogen fuel infrastructure program. The photovoltaic power will not only supply electric power and hydrogen fuel used by the recycling process, but it will also provide air-conditioning and other power used by the workers and their families. And while we are at it, let's provide the families with housing, schools, medical facilities, work uniforms, sewerage systems, and running water.

So what does all of this giveaway program cost? How about $30 million per location including a portion of one zeppelin transporter. This is our "fishing rod" for about 300 people per location, or 4.5 million people total. At $30 million for each of 15,000 locations, the grand total would be about $450 billion. This money would be provided by the countries participating in the program, and here is what they would get in return:

If we assume that twenty-seven thousand square miles of land is being abandoned worldwide each year, we could restore this land—over a twenty-year time frame—by the processes described above. Therefore, additional people could raise animals on the land and sell products and services to the recycle workers and their families. This would set up a mini—almost self-sustaining—economy as time passes.

Use of almost 90 square miles of photovoltaics will help bring costs of photovoltaic panels down to where they may even be cost-competitive with conventional power. As a result, more and more photovoltaic panels will be employed to provide power to a growing community that can afford to buy power. In addition, in the initial stages, 20.5 thousand megawatts of electricity would be produced by renewable fuel—the sun.

Since each location will be an entrepreneurial center, we can expect each location's size and profits to grow as more and more product manufacturers see the cost-savings advantages of the recycling system.

As the economy at each location grows, more and more people will become employed, and as a result, the original sponsors of the $450 billion will gain back their investment in tax revenue and a self-sufficient population.

Product manufacturers will design their new products with the recycling system in mind (e.g., using rust-resistant aluminum or titanium rather than steel). They will also gain a new market for specially designed goods—like small hydrogen-powered cars.

A detailed cost estimate for all of the above will be provided by me when, and if, this plan is seriously considered for implementation.

Oh yes, one other thing. Remember in the last subchapter how we were going to increase the amount of vegetation to absorb the greenhouse gas-carbon dioxide? Well, as we resuscitate the abandoned, degraded farmland, we are also increasing the amount of land that can absorb carbon dioxide. In fact, if we assume six thousand acres of restored land per recycle location, the result would be about 140,000 additional square miles of carbon dioxide-absorbing land.

4.5 The Future and Beyond

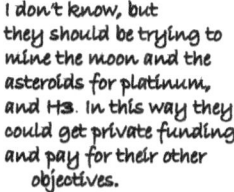

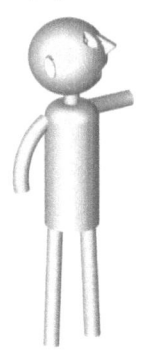

The year is 2050, and after many delays and funding cuts, the National Aeronautic and Space Administration (NASA) has successfully established a moon base and completed a manned trip to Mars. As it was after the first trip to the moon in 1969, NASA's funding is about to be cut again. No life has been found on Mars, and the public is asking why we spent all this money with so little to show for our effort.

The energy problems are getting worse on planet Earth, and the idea of space exploration is a last priority. In 2028, we had almost run out of oil in spite of small contributions from the Arctic National Wildlife Refuge (ANWR) and the Gulf of Mexico. And when the supergiant oil reservoirs in the Middle East had peaked in 2014, the price of gasoline skyrocketed. However, in spite of efforts to keep up with transportation fuel consumption by using biofuels, natural gas, oil sands, coal, and oil shale, the world oil consumption has continued to increase at a rate of 2 percent per year. This is continuing now, mainly because of the continued rapid industrialization of third world countries.

When the problem became unbearable, people began to accept the additional pollution and greenhouse gas emissions resulting from the conversion of oil sands and coal into transportation fuels. Fortunately, the polar ice caps did not melt enough to cause a worldwide devastation from rising sea levels. And because fission nuclear power and renewable energy sources have become more dominant for electric power generation it appears that fossil fuel usage will subside, and hydrogen produced from water and cellulostic ethanol will become the main source of energy for transportation. In addition, prototype fusion nuclear power plants are now feasible and safe, but are far from being economic.

Other problems have resulted from the world population increasing from 6.6 billion in 2008 to 9.2 billion now. What happened was that while the world's croplands were being used to make ethanol, a resulting massive food shortage led to significant increases in world poverty, hunger, and disease. This resulted because at the time ethanol was becoming popular, no one seemed to be paying much attention to the rate of cropland degradation, even though the signs were very clear. In fact, twenty-seven thousand square miles of cropland continues to be degraded each year. And as you might have guessed, the cost of food has skyrocketed.

You may have heard that back in 2008 that we had hundreds of years' supply of natural gas, oil sands, coal, and oil shale. Well, that shortsighted viewpoint did not consider that the 2 percent annual rate of worldwide consumption would continue. After all, we were going to use alternative fuels, conserve energy, and increase energy-consuming product efficiencies. What we didn't foresee was that the rapid industrialization of China, India, and other emerging countries would outpace any efforts to curtail it. The problem now is that the increased worldwide use of energy has almost depleted our known reserves of natural gas, oil sands, and coal. Oil shale is becoming more popular now, but we can't extract it fast enough to keep up with demand.

We have all come to realize now that hydrogen fuel is, and was, the way to go for transportation vehicles. It is safe to use, and it produces no atmospheric pollution or carbon dioxide. However, perhaps due to special interest groups, politics, misunderstandings, and/or fear of its use, it was relegated to a future "pie in the sky" solution when it should have been aggressively pursued. In the year 2008, we knew what the situation was, but we continued to put our heads in the sand. There should have been an

Apollo Moon type of project to build a hydrogen infrastructure. At that time, we could have used liquefied hydrogen to help overcome hydrogen's storage volume problem and developed our current carbon nanotube storage method along the way.

Hydrogen was originally derived from fossil fuels, such as natural gas, methane hydrates, and coal, because of cost. For obvious reasons, we now produce hydrogen from water using the wind to provide electric power for electrolysis. Although many hydrogen opponents used to say that too much electrical energy was used to extract hydrogen from water, what they did not appear to consider, or understand, was that the hydrogen energy would be used more efficiently in a fuel cell or hydrogen hybrid vehicle. In other words, hydrogen used with a fuel cell vehicle would double the mileage for the same amount of energy contained in gasoline. While it takes 55 kilowatt-hours to produce the equivalent amount of hydrogen gas energy as a gallon of gasoline (which has 36 kilowatt-hours of stored energy), its use in a fuel cell car would make it equivalent to 72 kilowatt-hours of gasoline energy. The result is not as dramatic with a hydrogen hybrid, but it would be 25 percent better than a gasoline hybrid, or 45 kilowatt-hours of effective energy and mileage.

In 2008, the presumed cost problems associated with hydrogen fuel cells was about to be solved, but the words "Hydrogen is decades away" rang throughout the legislative halls. However, there was a problem with fuel cells that needed to be addressed. And that problem was the platinum used for the fuel cell electrodes. If the U.S. Government had acted at that time, we wouldn't have the platinum resource problem that we have today. Fortunately, the hydrogen hybrid has helped to solve the platinum resource problem, but the need to further improve fuel efficiency with a hydrogen fuel cell has made the search for platinum much more important.

Back in 2008 when platinum was valued at more than $1,000 per ounce, a 60-kilowatt hydrogen fuel cell automobile would use about $450 worth of platinum. But because of the demand for fuel cell automobiles, the price of platinum has tripled in equivalent 2008 dollars. This increased value has caused NASA to redirect its efforts to determine if platinum exists on the moon. The theory being that some of the earth's near-surface metal deposits were caused by asteroid collisions, and the same could be true for the moon. And with the moon's reduced gravity, it is possible that asteroids containing platinum could be located close to the moon's surface. This theory is based upon a known asteroid impact that resulted in the Sudbury, Canada, nickel mine and other asteroid impacts that resulted in other mineral mines around the world.

Scientists believe that when the moon and earth were solidifying, certain elements combined with oxygen and floated to the surface. In contrast, elements—like gold and platinum—that did not react with oxygen tended to sink toward the center of these bodies. In fact, it is believed that there is enough gold near the center of the earth to cover the world's continents to a depth of about one foot. If this is true, then the same nuclear fusion process that formed the earth and moon elements could be true of asteroids and comets. And as some of these asteroids and comets impacted the moon, the gold and platinum elements would be deposited near the moon's surface because of its reduced gravity.

NASA's redirected effort to discover platinum could be the thing that will revive their fading program. In addition, the economics of fusion nuclear reactors could be improved by the use of another very expensive element that is known to be abundant on the moon. This element is He-3, or helium-3, a naturally occurring substance deposited on the moon's surface from solar winds. Currently, 2050, the cost of He-3 is about $5 million per pound. In contrast, platinum is worth only $5,000 per ounce. At present, a He-3 fusion reactor prototype is being developed, but to date, only small-scale experiments have been conducted. The advantage of a He-3 fusion reactor is that it can generate electricity directly using the released electrically charged proton when He-3 atoms are fused with other He-3 atoms. This means that the very expensive steam equipment—like steam turbines—would not be necessary to generate electricity. In addition, it doesn't require the massive containment vessel required for the conventional deuterium-tritium fusion process.

With He-3 priced at $5 million per pound, NASA could fund a limited space exploration program. However, the desire to make fusion nuclear more economical will cause NASA to bring He-3 back with minimum profit. On the other hand, mining He-3 could result in making the mining of platinum more practical.

Now here is where NASA and private enterprise could work together. In fact, they should have been working together years ago, instead of focusing on the search for life on Mars. If some of the money spent on the 2008 Mars project had been directed toward mining the moon, NASA would not be in the financial and public acceptance predicament that it is in today. In fact, the earth would probably not be in the energy predicament it is in now. Using the technology that it would have developed during the mining operation, it would have set the stage for Mars exploration and for exploring our closest star systems. Let's see how!

On second thought, I am not going to explain how. But, it could be the basis for an interesting science fiction book or movie—Hmmm. Here is a clue. Remember NASA's $305 billion solar collector described in subchapter 2.4? Well years ago NASA recommended a huge solar collector that included a one mile diameter rotating satellite that simulated earth's gravity on the inner surface. It was intended to be a self contained habitat, including an atmosphere, for humans to live in space. How about a place to process mined materials that are catapulted from the moon?

Chapter 5
Why Not

You can't help the impoverished by impoverishing the rich.

—Ronald Reagan

5.1 Git'er Done

A Sense of Urgency

Never in the history of mankind has the world faced a problem quite like the coming *transportation* fuel crisis. Yet so many people think that magically, something will be done to save us. Politically correct words like *conservation*, *efficiency*, and *ethanol/biofuels* are being bantered about by our government, the media, and oil company ads as if they were real solutions to the world's fuel problem. Are they not aware of how serious the coming fuel crisis is? Or do they know, but think that the people are incapable of understanding what is really going on.

Hopefully, after reading this book, you now understand what is going on. In subchapter 2.5, I did the arithmetic for a 2 percent per year increased worldwide fuel consumption rate and showed that the world's known fossil fuel reserves (crude oil, oil sands, oil shale, and coal) will be gone by the year 2067—assuming that the supply rate is able to keep up with demand and that all of these reserves are devoted to transportation fuel, not for electric power generation, heating, or industrial usage. I also estimated that conservation, energy efficiency, and ethanol will only help to prevent this from being 3 percent or more.

With regard to using natural gas as a transportation fuel, this could be an intermediate term solution to wean the United States off imported oil. The problem is that significant vehicle and infrastructure changes would be necessary. And, since the United States owns only 5 percent of the world's proven natural gas reserves, while Russia and the Middle East own 73 percent, the potential exists for future dependence on these countries if a hydrogen infrastructure is not developed in time. If a vehicle and infrastructure change is going to be made, why not leapfrog natural gas and develop a hydrogen infrastructure. In fact, we could use coal and natural gas to make hydrogen during the initial stages of vehicle and infrastructure development. Making a dramatic "urgent" vehicle and infrastructure change will be painful and the public may be reluctant to doing it twice. I don't believe that the same "sense of urgency" will happen the second time.

Since we do need to free ourselves from imported oil let's aggressively drill for more domestic oil and build coal to synthetic fuel plants with carbon dioxide sequestration. Realizing that these actions will take four or more years before we see the results, why not also implement shorter term solutions such as conservation, fuel efficiency, ethanol/biofuels, and begin the hydrogen proposals described in subchapters 3.5 and 3.6. Also, as previously mentioned, I believe that even though the aggressive domestic oil and coal to synthetic fuel programs will take four or more years to implement, the effect will be an immediate stabilization of oil and gasoline prices. This will happen because speculators will see a near future reduction in oil prices.

.

Don't be fooled by the skeptics who claim that fossil fuel reserves are understated, or that "peak" oil will not occur for another thirty to fifty years. All you have to do is look at the expected world transportation fuel consumption in 2050. At that time, the equivalent of today's Saudi oil field reserves would be depleted in about 3.5 years—and nobody expects to find another Saudi-sized super giant oil field. In 2050 world oil consumption terms, the recent 2006 oil discovery in the Gulf of Mexico would last 2.5 months. And that is assuming that the discovery turned out to be at the 15-billion-barrel upper-end estimate.

There are those who believe that the 2 percent exponential rate of fuel consumption can be reduced by telling people to conserve, to be more fuel-efficient, and to use ethanol. Yes, we, in the industrialized countries of the world, can do the arithmetic and show that we can do better, but the industrialized world is not the problem. *In fact, arguing against conservation and energy efficiency is like arguing against motherhood and apple pie—make no mistake about it, I am all for it.* The problem is that the current, non-fuel-using world population is growing rapidly and becoming more industrialized. For them, it is not a question of conserving, being more energy efficient, or growing ethanol. Rather, it is a question of denying or giving them access to fuel and other resources that will help better their lives. *Are we saying (or thinking) that only a small percentage of the world's population deserve to use the world's nonrenewable natural resources?*

Keep in mind that 300 million people in the United States consume 25 percent of the world's oil supply, and there are currently 6,600 million people on this planet (i.e., 4 percent of the people consume 25 percent of the oil). And as previously stated, it doesn't make sense to impoverish the rich to help the impoverished. With an inexhaustible supply of hydrogen fuel available, it further doesn't make sense to put off implementing the inevitable solution until tomorrow when it can, and should, be started today.

Oh, you think that supply-and-demand pricing, fuel taxes, and rationing will come to the rescue and reduce the 2 percent exponential rate of world fuel consumption. OK, you are right, *but what are the unintended consequences* of these actions? How about economic chaos in the industrialized countries, denial of a better life for the people in the nonindustrialized countries, and continued exponential world population growth? When you look at population growth, relative to industrialization, you will find that as a country becomes more industrialized, the more unlikely it is for its people to have large families. Japan, for instance, currently has a negative population growth rate.

Or you might be one of those optimists who believe that fuel prices will remain close to where they are now because we will keep finding more oil (and other fossil fuels) and increase the use of ethanol. Maybe so, but how much do we need to find and grow to keep up with the world's 2 percent exponential consumption rate? I proved the magnitude of the problem with the previously mentioned oil discovery in the Gulf of Mexico. And with regard to ethanol, just look at the expected world population growth over the next forty-plus years (2.7 billion additional people—more than India and China combined) and then explain how all those people will be fed, while at the same time depleting the diminishingly available agricultural land to make ethanol.

Except for renewable energy sources, breeder fission, and/or fusion nuclear power and hydrogen fuel, all of the above-mentioned technologies and management techniques will fail in the long run. It is a matter of how dedicated, focused, and innovative we are in implementing renewable energy sources, fusion nuclear electric energy, and hydrogen transportation fuel that will save the day.

And what are the consequences of continuing to use up all the world's known fossil reserves and not focusing and accelerating the above-mentioned alternatives? First of all, in a free market, all of the various types of fossil reserves will be extracted as fast as possible in an effort to keep supply-and-demand pricing from causing worldwide economic chaos. Second, in response to higher fuel prices caused by supply and demand, almost everyone that has land will grow ethanol rather than food and livestock—thus causing higher food prices and worldwide starvation. Third, we will be spewing carbon dioxide and pollutants into the earth's atmosphere at an unprecedented rate, causing who knows what to happen.

There should be no doubt that if we continue to use fossil fuels, as we have been, without a dramatic focused and accelerated shift to renewable energy, hydrogen fuel, and fusion and/or breeder fission nuclear power, there will come a time when the world will run out—assuming that we are able to survive until that time without having a total economic breakdown or a global war.

The current rate that we are implementing renewable energy sources and hydrogen fuel for transportation is tantamount to total incompetence and ignorance on the part of our global leaders. There needs to be "a sense of urgency" like the Apollo Moon Program to develop a wind-, solar-, and water-based hydrogen fuel infrastructure (possibly coupled with a to-be-determined percentage of hydrogen made from ethanol) now!

The pessimists who can't seem to find enough reasons for not using hydrogen fuel are not using common sense. *We will use hydrogen fuel in the future, and the only question is when.* While new processing and storage technologies may happen in the future, we can't afford to wait. Technologies exist now to implement a hydrogen fuel infrastructure, and no breakthrough technology is needed. When new technologies do come along, we will adapt.

The Hydrogen Plan

During the early stages of the hydrogen fuel program, I would pursue as many alternative gasoline equivalent fuel sources as practical—primarily drilling for more domestic oil and making synthetic fuel from coal. The idea being that as long as hydrogen is the centerpiece of a larger plan, that these other alternatives would provide a bridge, and accelerate making the United States less dependent on oil imports. Natural gas should be part of the plan, if it is used to make hydrogen during the initial infrastructure development. To develop a parallel natural gas vehicle and infrastructure would, in my opinion, dilute and confuse an all out effort to develop a hydrogen infrastructure.

The key factor in my proposed hydrogen program is a commitment to completing the entire hydrogen fuel infrastructure in a short period of time—with the *urgency* of a Manhattan Project or Apollo Moon Program. This is the most important requirement because robotically equipped advanced mass production factories are required to keep costs down. Probably the second most important factor is to develop the 20-megawatt wind-driven electric generator—including the large-scale zeppelin—since my analysis shows that about 60 percent of the cost to produce and deliver hydrogen fuel is wind-generated electric power. While the currently designed Magenn floating air rotor does potentially provide the underpinnings for low-cost hydrogen fuel, it is probably not the best solution. Since I am recommending that the program be initiated with a two-year feasibility study, I would evaluate both the 5-megawatt Magenn approach and the "breakthrough" 20-megawatt approach (including the 10-megawatt air rotor) and make a determination when the study and test phase is completed.

In addition, the supporting equipment, related to the production and distribution of hydrogen fuel, should be developed—including alternatives to standardized equipment. This includes electrolysis equipment, liquefaction equipment, superconducting and HVDC electric transmission, hydrogen pipeline equipment, liquid hydrogen storage tanks, high-capacity liquid hydrogen delivery vehicles, and robotic service station equipment.

If we were able to improve upon the amount of energy required to generate and liquefy hydrogen from water, it could result in a significant savings. And remember this, all of the hydrogen-producing equipment presented in subchapter 3.1 would be paid off in ten years, which means that if new and better technologies are developed, the old equipment can be replaced with new equipment.

Another parallel effort would be to get the hydrogen infrastructure started by producing hydrogen at existing coal-fired power plants and sequestering the carbon dioxide. The low cost of doing this would pay for the sequestering equipment and retrofitting vehicles. Specifically, if all of the nation's diesel-powered equipment (trucks, tractor-trailers, trains, etc.) were to be retrofitted with supplemental hydrogen fuel tanks that allow a 60 percent hydrogen and 40 percent diesel fuel blend, we would save 63 percent of our current Middle East oil imports. At the same time, the diesel equipment would use less fuel, gain 20 percent more power, and emit less pollution.

Retrofit kits could also be developed to equip existing vehicles with dual fuel tanks—hydrogen and gasoline. By providing incentives to automobile manufactures, service mechanics, gas stations, and vehicle owners, a hydrogen fuel infrastructure could emerge before the hydrogen, from wind and water, infrastructure begins. This program can also help to speed up the elimination of Middle East oil imports and can piggyback on the hydrogen fuel produced for the above diesel program.

With regard to the Ballard Company's effort to develop a suitable low-cost hydrogen fuel cell platform, I would make the development of an alternative to platinum anode and cathode coating a priority. In parallel, I would make platinum a strategic material and begin large-scale, low-cost, platinum mining operations—perhaps using my proposed zeppelin technology.

During the two-year feasibility study, I would expand and accelerate the research currently being done to develop new hydrogen fuel tank technologies. While liquefied hydrogen can be safely handled, I would like to see a gaseous hydrogen fuel tank that meets or exceeds the Department of Energy (DOE) goals. In parallel with this effort, I would like to see a lightweight free-form liquefied hydrogen fuel tank developed that includes a twelve-day dormancy period before "boil off" and a "boil off" capture system.

Details of my proposed hydrogen fuel implementation plan will be disclosed when, and if, the proposals in this book are accepted. The plan includes proposed participants, quantities and types of mass-produced equipment, proposed government and private enterprise expenditures, and a timeline. The timeline begins with the initiation of a reduced-scale hydrogen fuel infrastructure (e.g., like the one described in subchapters 3.5 and 3.6), while a two-year feasibility program is being conducted. Following the feasibility program, a period of factory construction and product "debugging" is envisioned in a time frame that minimizes in process engineering changes once full production has started. Once the manufacturing and product installation processes are up and running, the implementation phase should go very quickly. However, the proposed mass production cost benefit requires large quantities of manufactured products to be produced, which means that without a commitment to a major portion of the plan, the economic aspects may fail—the urgency of a Manhattan Project or Apollo Moon Program is needed.

With regard to achieving the hydrogen fuel price goals, I would recommend a course of action that presumes dramatic price reductions for hydrogen-producing equipment. Take, for example, the hydrogen electrolysis equipment, which has a current non-mass-produced price—from one manufacturer—of three times more than what I have stated in my economic analysis. In this instance, I would challenge several electrolysis equipment manufacturers to propose a mass production plan that will achieve a 70 percent price reduction. Since we will need 95,000 electrolysis machines to produce 100 percent of our current fuel consumption, the total sales would be $950 billion at $10 million per machine. This should create enough incentive for three or more companies to participate in the challenge; however, there would need to be penalties for those who are unable to reach the price objective. These penalties, aside from the embarrassment of failure, should include reduced profits and reduced future sales. Conversely, an incentive plan should be established for the companies who exceed their objective.

Implementation of the Plan

Prior to initiating the feasibility program, the President of the United States needs to announce, to the public, this country's intention to move to a hydrogen fuel economy within sixteen years, with a major first step (30 percent) being completed in nine years.

The above announcement is needed so that the public, other countries, and companies dealing in oil and other alternatives are forewarned so that they can make appropriate economic decisions that affect them.

The hydrogen fuel project should be announced in a manner similar to President Kennedy's speech that said, "We will go to the moon in this decade." Like the Manhattan Project, Liberty ships, and the mission to go to the moon, the hydrogen fuel project should carry the same sense of urgency and commitment.

The public needs to know what "peak" oil is and what it could mean to our entire way of life in the United States and to the world. They need to know that conservation, fuel efficiency, ethanol, oil sands, oil shale, synthetic fuel from coal, natural gas, methane hydrates, and conventional "fission" nuclear power, *while initially desirable*, are only temporary measures. They need to know that these measures, even in an all-out effort, will not meet demand and will have a detrimental air pollution effect and the potential effect of carbon dioxide emissions. They would also need to know that the United States is committed to an *"all-out effort"* to reach a near-term hydrogen fuel infrastructure. And because of this accelerated and focused effort, near-term "peak" oil-mitigating technologies and management techniques should be applied until the hydrogen fuel infrastructure is in place.

If explained in the foregoing manner, a large majority of people will respond favorably. They will know and understand that the "peak" oil crisis is real and that their government is doing the right thing, in the near term and long term, to protect them against overburdening fuel prices and usage restrictions. When fully understood the people will consider short-term fuel conservation and efficiency to be an act of patriotism and a legacy to their grandchildren—and not a political subterfuge.

Since previously proposed solutions for fuel and energy independence vary all over the map, the average person doesn't know who or what to believe. Some continue to think that "magically," everything will be OK. Therefore, it is of the utmost importance for the United States Government to clarify what is happening and propose a real solution for which almost everyone will agree. Not to do so would be unjustifiably irresponsible.

The Future

While reading this book, you may have wondered what the future world would look like with so much land and ocean area being occupied by the wind and solar energy to sustain the future population. Well, to help make this projection, let's make the following assumptions and do the arithmetic:

1. The world population peaks at 11 billion people due to industrialization and improved standards of living—derived primarily from a worldwide hydrogen fuel infrastructure.
2. Each of the 11 billion people consumes hydrogen fuel at the per capita rate used by the United States today—467 million kilograms of hydrogen fuel per day for a population of 300 million people.
3. In a worst-case scenario for wind and solar energy density, we will assume 18 megawatts of electric energy output per square mile of land or ocean (equal to using the current 5-megawatt propeller-and-pedestal wind turbine technology).
4. Seventy percent of future hydrogen production is ocean-based; therefore, 30 percent is land-based (except in the case of ethanol being used to supplement the production of hydrogen).

With these assumptions, about 1 percent of the world's ocean area would be required, and 1 percent of the world's land area would be required. And this does not account for the improved vehicle efficiency resulting from the use of hydrogen fuel—especially a 100 percent improvement from using hydrogen fuel cells.

I do not believe that any person in the world should be denied the benefits that come from our modern industrial and medical achievements. And the hydrogen fuel scenario, presented in this book, provides the centerpiece for allowing *all* of the world's people to live in prosperity.

If economical fusion nuclear power plants become part of this future, the production of hydrogen could be accomplished by separating hydrogen and oxygen from water at high temperature—rather than by electrolysis.

Things That Never Were

The "things that never were," discussed in subchapters 4.1 through 4.5, were presented as food for thought. They are inventions from my mind that could be a spin-off from the hydrogen fuel development program and are not meant to be any more than ideas to consider—you can be the judge. These inventions from my mind are as follows:

1. Generating "green" electric energy, from the wind, at a cost that is about the same as conventional fossil-fired or nuclear power plants—the 20-megawatt wind-activated electric power generator
2. Incorporating low-cost, semi-self-powered, large-scale zeppelin technology into a new way of doing things. These things include people and cargo transportation, disaster relief, low-cost mining operations, low-cost aqueducts to provide water to drought-stricken lands, and providing affordable housing
3. Transforming deserts into atmospheric carbon dioxide removal systems that also supply hydrogen fuel
4. Renewing degraded farmland and providing worldwide recycling of nonrenewable resources
5. Mining the moon and asteroids for platinum and helium-3

Low-Cost Wind Power

Using the Magenn floating air rotor as the platform for future electric power generation, a breakthrough technology was proposed. This breakthrough technology positions the floating air rotor at ground or ocean level to provide all weather protection and enhanced performance through wind concentrators and blade positioning. As a result, it has been shown that output capacity can increase more than eightfold with a doubling in size. As a result, 20-megawatt generators were proposed that produce electricity at about half the price of conventional propeller-and-pedestal wind turbines and about the same price as conventional fossil and nuclear power generation. The 20-megawatt generator also consumes only about three times the deployment area required for conventional power generation. And when devoted to hydrogen production and supplemental electric power, their intermittent performance is not an issue.

Self-Powered Zeppelin Technology

One of the technologies used to keep the 20-megawatt generator costs down was the use of a large-scale core zeppelin. This technology was proposed as a means of transporting the rotor portion of 20-megawatt generator from its factory to its deployment site. Because the zeppelin would be mass-produced at the 20-megawatt generator's factory, its cost would be relatively low, thus offering other possibilities for low-cost air transportation. Combining its low cost with low-cost mass-produced thin film photovoltaics, the large surface area of the zeppelin could generate a significant amount of electric power. Transforming this electric power into hydrogen fuel allows the zeppelin to be self—or semi-self-powered through the use of fuel cell-driven electric motors. Although the zeppelin's speed capability would be low if it relied only upon its self-generated fuel, there are many uses that don't require high speed. In cases where higher speed is required, additional hydrogen fuel tanks and higher horsepower motors can be added.

Transforming Desert Lands

One of the uses for the large-scale zeppelin is the transportation of aqueducts, soil, and fertilizer to currently unused desert lands where cellulostic ethanol crops can be grown. Transforming these crops into hydrogen fuel—made from ethanol—required a significant amount of energy that would be derived from the 20-megawatt generators. Thus, by sequestering the carbon dioxide produced during the transformation processes, a cycle of plant growth and subsequent processing can reduce atmospheric pollution and carbon dioxide levels—a "net minus" effect. Besides providing a better standard of living for millions of people, the hydrogen-from-ethanol scenario may be the most economic approach to generating and transporting hydrogen fuel. However, because of land limitations, the amount of hydrogen fuel may be limited to about 40 percent for the United States.

Worldwide Recycling

Again, using the zeppelin technology and the development of low-cost, high-efficiency photovoltaics (resulting from the hydrogen-producing equipment mass production program), an innovative approach to worldwide recycling was proposed. In addition to the recycling of nonrenewable materials, this innovative approach had a potentially major side benefit. This side benefit was to restore degraded farmland and provide a better standard of living for billions of impoverished people around the world. An additional side benefit is the possibility of opening up new worldwide markets and economic growth. As a result of this economic growth, there is reason to believe that the exponential population growth will be significantly reduced. A reduced population growth would lead to a better standard of living for everyone and a stabilized use of nonrenewable resources. And because hydrogen fuel and wind power will be in abundant supply for all of the world's population, there will not be an energy crisis.

Mining the Moon

A case was made for redirecting NASA's space program from trying to find life on Mars and terraforming Mars for future earth inhabitants to mining the moon and asteroids for the future betterment of mankind. Besides the possibility of finding platinum on the moon to support the use of hydrogen fuel cell vehicles, the discovery of helium-3 (He-3) in the moon's soil (regolith) opens the door to a much more economical means of producing fusion nuclear power. By redirecting its efforts to mining operations, NASA could obtain private funding that would allow a much more productive space program without the need for taxpayer funding. By catapulting regolith, from the moon, for processing in a large robotically self-replicating, gravity-simulating (by rotation) satellite habitat, a number of side benefits would be derived. Among these side benefits would be the fulfillment of NASA's other objectives, which might even include a living space for the earth's future inhabitants and travel to our next solar system.

5.2 The Daffodil Principle

They were planted by a woman who proved that one person can make a big difference

Who planted all those daffodils?

As a conclusion to this book, I would like to relate to you a story about the *daffodil principle* and how the daffodil principle applies to the future of hydrogen fuel and to the betterment of mankind. *It is a story—authoress unknown—told by a woman who has a daughter and two grandchildren. Here is that story:*

Several times my daughter had telephoned to say, "Mom, you must come to see the daffodils before they are over." I wanted to go, but it was a two-hour drive from Laguna to Lake Arrowhead. "I will come next Tuesday," I promised, a little reluctantly, on her third call.

Next Tuesday dawned cold and rainy. Still, I had promised, and reluctantly I drove there. When I finally walked into Carolyn's house, I was welcomed by the joyful sounds of happy children. I delightedly hugged and greeted my grandchildren.

"Forget the daffodils, Carolyn! The road is invisible in these clouds and fog. And there is nothing in the world, except you and these children, that I want to see badly enough to drive another inch!"

My daughter smiled calmly and said, "We drive in this all the time, Mom." "Well, you won't get me back on the road until it clears, and then I'm heading home!" I assured her. "But first we're going to see the daffodils. It's just a few blocks," Carolyn said. "I'll drive. I'm used to this." "Carolyn," I said sternly, "please turn around." "It's all right, Mother, I promise. You will never forgive yourself if you miss this experience."

After about twenty minutes, we turned onto a small gravel road and saw a small church. On the far side of the church, I saw a hand-lettered sign with an arrow that read, "Daffodil Garden." We got out of the car, each took a child's hand, and I followed Carolyn down the path. Then, as we turned the corner, I looked up and gasped. Before me lay the most glorious sight.

It looked as though someone had taken a great vat of gold and poured it over the mountain peak and its surrounding slopes. The flowers were planted in majestic, swirling patterns, great ribbons of and swaths in deep orange, creamy white, lemon yellow, salmon pink, and saffron and butter yellow. Each different-colored variety was planted in large groups so that it swirled and flowed like its own river with its own unique hue. There were five acres of flowers.

"Who did this?" I asked Carolyn. "Just one woman," Carolyn answered. "She lives on the property. That's her home." Carolyn pointed to a well-kept A-frame house, small and modestly sitting in the midst of all that glory. We walked up to the house.

On the patio, we saw a poster. "Answers to the Questions I Know You Are Asking" was the headline. The first answer was a simple one. "50,000 bulbs," it read. The second answer was "One at a time, by one woman. Two hands, two feet, and one brain." The third answer was "Began in 1958."

For me, that moment was a life-changing experience. I thought of this woman whom I had never met, who almost fifty years before had begun, one bulb at a time, to bring her vision of beauty and joy to an obscure mountaintop. Planting one bulb at a time, year after year, this unknown woman had forever changed the world in which she lived. One day at a time, she had created something of extraordinary magnificence, beauty, and inspiration. The principle her daffodil garden taught is one of the greatest principles of celebration.

That is learning to move toward our goals and desires one step at a time—often just one baby step at a time—and learning to love the doing, learning to use the accumulation of time. When we multiply tiny pieces of time with small increments of daily effort, we too will find we can accomplish magnificent things. We can change the world.

"It makes me sad in a way," I admitted to Carolyn. "What I might have accomplished if I had thought of a wonderful goal thirty-five years ago and had worked away at it 'one bulb at a time' through all those years? Just think what I might have been able to achieve!"

My daughter summed up the message of the day in her usual direct way. "Start tomorrow," she said. She was right. It's so pointless to think of the lost hours of yesterdays. The way to make learning a lesson of celebration instead of a cause for regret is to only ask, "How can I put this to use today?"

Use the daffodil principle. Stop waiting . . .

Until your car or home is paid off

Until you get a new car or home

Until your kids leave the house

Until you go back to school

Until you finish school

Until you clean the house

Until you organize the garage

Until you clean off your desk

Until you lose 10 pounds

Until you gain 10 pounds

Until you get married

Until you have kids

Until you retire

Until summer

Until spring

Until winter

Until fall

Until you die . . .

There is no better time than right now to be happy.

Happiness is a journey, not a destination.

So work like you don't need the money.

Love like you have never been hurt, and dance like no one's watching.

If you want to brighten someone's day, tell them about the daffodil principle.

The Point of the Story

Well, what does the daffodil principle have to do with hydrogen fuel?

It's about changing the world, one step at a time, in a way that will help ensure a better life for everyone. If each reader of this book began today to spread the word, to all they come into contact, of the necessity for changing from gasoline to hydrogen fuel, it can potentially make a big difference. There will be a groundswell that will not be ignored. Questions will be asked at town meetings with elected representatives. Some will e-mail their Congressperson or Senator. Others will appear on television and radio newscasts and interviews. Letters to newspaper and magazine editors will ask questions and demand answers.

Like the woman in the story, if we all started to plant the hydrogen fuel story equivalent of daffodil bulbs, we will change the world for the better. Instead of one person planting bulbs for fifty years, we could plant the fifty-year equivalent of daffodil bulbs in just one year.

In the first chapter of this book, I expressed concern for my grandchildren's future because of the way that world is currently headed. And this book is intended to be a legacy to them. But on a larger scale, almost everyone in the world can have their lives changed, for the better, by what we all do here and now. In fact, it may be the starting point for the eradication of worldwide poverty, hunger, and disease.

As previously mentioned, the world population is increasing rapidly, and more and more of the world's people are making the transition from poverty to a greater degree of affluence. This affluence is coming as a result of global industrialization. Almost everyone would like to raise their standard of living, and probably more importantly, they would like to raise the future standard of living of their children and grandchildren.

Growing Pains

China is where we find an example of the growing pains that come from the people who are seeking a better life. Uncontrolled growth has led to substantial air and water pollution. As a result, disease and sickness is increasing at an alarming rate. There is an accelerating disparity between those who are participating in the industrialized growth and those who are not. Lack of water in the northern parts of China has caused the government to begin a fifty-year program to construct a canal to bring the freshwater-rich southern water to the north. However, much of the south's water has now become contaminated. The requirement to provide China's 1.3 billion people, and its new factories, with electric power has led to building more and more coal-fired electric power plants—about 2,200 new coal plants are scheduled between now and the year 2030. And the coal used contains a high percentage of sulfur. Lack of smokestack emission controls has led to China having five of the top ten world cities with the greatest amount of air pollution. Towns and villages are being uprooted to allow for the industrial spread, and unemployment is at a very high level. Many of China's citizens have to leave their families for long periods of time to seek work at long distances from their homes.

The driving force for this worker migration is their willingness to accept low wages rather than little or no wages. At present, the low-wage incentive is causing companies around the world to build manufacturing plants in China's industrial parks and importing the low-cost manufactured goods back to their country. For example, China currently exports $264 billion in manufactured goods per year to the United States while exporting only $50 billion. One result of this imbalance is the enormous number of containers coming into U.S. ports with an enormous number of empty containers being returned back to China.

The Hydrogen Fuel Infrastructure

What does this China example have to do with the daffodil principle or building a hydrogen fuel infrastructure in the United States?

To answer this question, we need to begin by restating that it should be obvious that China and other emerging nations will be driving the exponential demand for transportation fuel in the future (hopefully at less than 2 percent per year). This is because the United States, and other industrialized countries, are sending large sums of money to China and providing China with manufacturing expertise that will promote entrepreneurship. As wages increase and entrepreneurship develops in China, the people will be able to afford more and more self-made and imported goods—including oil (or hydrogen fuel) and robotically produced manufactured goods—made in the United States. China's entrepreneurs will build factories near where people live rather than at central locations, and companies like Wal-Mart will build more and more Chinese-style stores around the country (currently Wal-Mart has five stores in China—and plans to add many more). Tax revenues from this economic growth will be used to clean up the polluted air and water and extend a network of sewage and wastewater treatment facilities. Health care will be improved, and disease prevention will be emphasized. Highways will be extended throughout China, the south-to-north canal will progress, and advanced waste disposal facilities will be implemented. It all takes time.

In support of the improvements that can occur in China, we will assume that the United States is developing a hydrogen fuel infrastructure as prescribed in this book. As a result, the United States economy will grow and prosper. The advanced manufacturing, robotic, and recycling methods that will be developed will create a new manufacturing base that does not rely upon low-cost labor. The wind and solar energy-producing methods that will be developed will reduce the cost of clean electric power generation. These new power-generating methods will also provide an alternative to China's planned 2,200 coal-fired power plants.

Hydrogen-powered vehicles and advanced manufacturing methods will revitalize the automotive industry in the United States. Superconducting electric power transmission methods will reduce electric transmission costs, eliminate the need for unsightly overhead transmission lines, and help to establish an advanced national electric grid. Food will become a major U.S. export as a result of increased world population. Air pollution and carbon dioxide emissions will steadily decrease. And the empty containers that are currently being returned to China will be filled with food and robotically made manufactured goods.

Because of the success of the hydrogen fuel project in the United States, other countries will follow—including China. Initially the United States will export hydrogen-producing equipment and liquefied hydrogen—via catamaran tankers (like the one described in subchapter 3.7). Later, these countries will become self-sufficient in the production of hydrogen fuel, which will stimulate their economies. The United States will also export technologies and equipment that will clean up air and water pollution in the newly industrialized countries around the world. And eventually, fossil-fueled electric power plants will be replaced by clean and safe breeder fission and/or fusion nuclear power, which will be supplemented by wind and solar electric power—or vice versa.

In time, each country—like China—will become more and more self-sufficient and generate economic growth within their own country. This economic growth will produce a steady improvement in people's lives. The disparity between the people who have, and those that don't, will be lessened within their countries, and the disparity between countries will be lessened. Each country will share its natural resources in an open market, and because full-scale product recycling will predominate, no country will be dependent on another for the earth's nonrenewable ingredients—as it is today with oil. And because almost everyone will be participating in the current and future industrial achievements, the desire for war will be lessened, and the world population growth will be brought into equilibrium. Because of this better standard of living, a greater number of the world's people will be better-educated and more able to participate in the development of new technologies and cures for life-threatening diseases.

The stimulus for this possible utopian future could be the focused and *urgent* development of a hydrogen fuel infrastructure by the United States Government. For the United States to be a world leader, it needs to use its immense talents and God-given resources to not only help its own people, but to help all of the world's people. Then, and only then, will the United States be looked upon by the world with respect, admiration, and gratitude.

All this could result because you, and others like you, started planting daffodil (hydrogen) bulbs. This book is my daffodil bulb.

Appendix

A Cure for Insomnia

This appendix may contain more than you need to know, but not enough for some. And if you have insomnia, it could put you to sleep. If you are technically inclined, the following discussion might provide some answers to questions that you may have developed while reading this book; however, this is still the short version of a very complex subject. If you are still not satisfied with these details, it may be necessary for you to refer to the information provided in the list of references. If still not satisfied, you may need to do your own research.

In this appendix, we are going to examine the basis for my cost estimate claims made in this book. In effect, it is an elaboration of information provided in subchapter 3.1, except for the subject of wind power, which was discussed in some detail in subchapter 4.1. Also, for those of you who like to deal in kilometers rather than miles, we will include the metric form of measurement. So here are the subjects that will be covered as they pertain to hydrogen fuel cost and infrastructure:

- *Advanced* Mass Production
- Solar Power
- Electrolysis
- Superconductivity
- Hydrogen Gas Transportation
- Liquefaction
- Storage
- Delivery

Advanced Mass Production

Since *advanced* mass production is the crux of how our low-cost energy scenario can be achieved, it is necessary to explain this concept in enough detail so that you can attach some sense of reality to the cost estimates proposed.

Remember the Model T Ford that Henry Ford introduced at a low-enough cost so that the average person could afford one—including his employees (he also paid them a relatively high wage). His plan was to mass-produce, via an assembly line, a car that costs much less than one that was individually built by hand.

Henry Ford vertically integrated about 70 percent of the parts and subassemblies (i.e., he made 70 percent of his own parts and subassemblies) of the Model T into his factory to keep costs as low as possible. To further reduce costs, as the story goes, Henry asked one of his suppliers to ship their parts in a box that later became the Model T running boards.

To keep costs as low as possible, Ford Motor Company was heavily vertically integrated for a long time before it became clear that they could not keep up with the technology changes that were occurring. When you have an evolving product that is undergoing technological changes, it is sometimes better to let the free market work on your behalf. For instance, the design of hydraulic systems for power steering and brakes, automatic transmissions, and other changes were best left to people who specialized in the development and manufacture of these devices.

As the automobile became more complex with more and more part or subassembly suppliers, the outside purchasing of parts became a costly problem. The first method in dealing with this problem was to have multiple suppliers who would bid their costs down. As a result, since most suppliers were not sure of their production requirements, they stored large quantities of inventory to make them responsive to changing demands (and this was reflected in the price). Furthermore, these suppliers were hesitant to invest in high-volume tooling and robotics that further raised the potential cost of their product. In addition, the automobile manufacturer had so many models, and options on each model, that purchasing agents often would pay high premium prices for some parts to be available in time for assembly (further increasing costs). And again, to avoid part shortages, the automobile manufacturer housed large inventories that in some cases were made obsolete because of design changes. And all of these storehouses of inventory had to be—in some cases—heated, lighted, maintained, and possibly air-conditioned (further driving up costs).

When the Japanese automobile manufactures introduced *just-in-time* manufacturing, they were able to produce lower-priced, higher-quality automobiles with more options than their counterparts in the United States. Ironically, it was an American, Edwards Deming, who showed the Japanese how to do it (his words fell on deaf ears in the United States—sound familiar?). Without getting into this too deeply, let's try to summarize what Deming proposed:

1. Form alliances with fewer suppliers and have parts or subassemblies delivered to the automobile assembly plant one or two days before they were scheduled for use (i.e., just-in-time).
2. Have part and subassembly suppliers require that all of their suppliers follow the just-in-time methods.
3. Use statistical process control to ensure that parts and processes were duplicated precisely every time (no quality inspection needed after the part, process, or subassembly is completed).
4. Each worker on a part, process, or subassembly inspects their own work and stops the assembly line if there is an instance of nonconformity.
5. When the assembly line is stopped the workers, engineers, purchasing agents, and possibly the part or subassembly supplier have a meeting to determine the root cause of the nonconformity—and make appropriate corrections (therefore, no need for rework of assemblies after they came off the line—therefore, more factory space and less labor).
6. Minimize process-set up time through the use of quick change dies.
7. Use the available factory space, from low inventory, and no final inspection and rework, to expand the production capability of the factory—thus reducing costs.

There is more to it than is being described here, but with this background, you may be able to understand how an even better and less expensive way to mass-produce products can be achieved through advanced mass production.

However, one final thought before we discuss how *advanced* mass production works. Let's discuss how some costs of materials, parts, and subassemblies are higher than they should be:

1. Even experienced engineers sometimes do not take advantage of standardization (one example of this is that automotive wheel bearings are mass-produced in such great quantities that you can buy them for a very low cost—by selecting a bearing that is one size larger or smaller, it is possible to inadvertently pay as much as 500 percent more).

2. By not reviewing the product "cost roll up" on a regular basis, it is possible to inadvertently pay very high costs (one actual case involved a $1 clear plastic shield for protection of an electronic printed circuit board—it turns out that the purchasing department was paying $50 for this part because that is what was paid to make a single prototype—the product went into production without renegotiating the cost).

3. The price you pay for a product generally includes the amortized cost of specialized equipment used to make the product. If the product is currently being sold to another user, it may be possible to get the same product at a much lower price, since it would then be incremental sales to the supplier.

4. Using the example in item 2, there is a lot that can be done to reduce purchased product costs through price negotiation. In fact, it is where most product cost reduction comes from.

5. Since purchased part or subassembly costs are strongly tied to volume, it makes sense to look at commonality with other products made by the same manufacturing company—many times one part of the same company doesn't know what another part of the company is doing.

Well, let's stop here; there are more examples, but as you can see, there are lots of things that make up a product cost that need to be considered when setting the stage for my proposed *advanced* mass production process.

OK, here is a summary:

1. Make detailed bills of material (back to the raw material stage), including all parts and subassemblies for all of the above components.
2. Sort the above bills of materials to determine major mineral groups—about 70 percent to be produced from on-site delivered raw materials (iron ore, copper ore, bauxite for aluminum, ethane and propane for plastics, etc.).
3. Further sort the bills of materials for common or similar parts—like motors, bearings, etc.
4. Redesign products, if necessary, to use common, similar, or standardized parts.
5. Determine the percentage of parts that will be made at a single on-site factory versus the parts that will be purchased.
6. Encourage specialized manufacturers to set up on-site facilities to make their materials, part, or subassembly and negotiate a required price.
7. Develop single-source supplier relationships for off-site materials, part, or subassembly manufacturing and negotiate a required price.
8. Develop a preliminary design for the on-site factories and materials supply systems.
9. Determine the electric power requirements to operate the factory and negotiate with the electric power companies a long-term kilowatt-hour price for electricity on the assumption that when air rotors are producing electric power, some of the power will be sold at a low cost to the grid system—this low-cost power is especially needed for aluminum production.
10. Attract a labor force to move to the factory area based upon about $50 per hour wages, with health and retirement benefits (also attract a competent engineering and management team).
11. Form an on-site team of architects, manufacturing engineers, chemical process engineers, and design engineers from specialized supplier companies in a cooperative effort to design the factories, tooling, robotics, and product delivery systems; synergize the product mix (common motors, bearings, etc.); design common feeder factories or areas within a factory (raw material process centers, a machining center, a die-casting center, a wire-drawing center,

etc.), and product-part manufacturing, and assembly/subassembly lines that maximize the use of hard tooling, fixtures, and robotics.

12. Construct the factories and order enough materials and equipment to begin trial product runs.

13. Perform trial product runs until almost *all* design and manufacturing issues are resolved (including input from factory worker personnel)—*consistent manufacturing of an unchanged product—with no optional equipment changes—for a long period of time is the most important ingredient in the low-cost advanced mass production process.*

14. Begin the advanced mass production manufacturing process and terminate all engineering change orders—except for emergencies and those that are approved by a joint manufacturing team.

For some people reading this book, this *advanced* mass production description has too much boring detail; for others, it's not enough. But those who understand this process know full well the importance of items 13 and 14.

One final thought: if *advanced* mass production is proven to be successful, there could be another future benefit. For instance, with a growing worldwide middle class, the demand for automobiles and other manufactured goods will increase. Therefore, it may be possible to produce an automobile, dishwasher, or other product at such a low cost that the United States could be a net exporter to China, India, and other emerging countries. The idea being that a proven design, basic model could be produced without change for as many as ten years or more. The key is using a design that doesn't undergo continuous design change and require options. For instance, if a quality automobile could be mass-produced for a *price* of $5,000 instead of $20,000, the world will buy it. In fact, this design consistency approach to mature products could lead to a revolution in recycling technology as more and more of the earth's resources are consumed by the industrialization of a growing world population. A disassembly line, for worn-out assemblies, running beside a new product assembly line, may make a lot of sense in the future. In fact, designing for robotic disassembly may well be a cost-effective criteria for all new products. Keep in mind that most new products are made from materials that are unique to that product. In other words, aluminum from pop cans cannot be used directly for the manufacture of structural-grade aluminum parts.

Solar Power

Solar energy is the conversion of sunlight into some form of energy. But the average person is probably confused about how this is done and how much of the sun's energy can be obtained from a given surface area. In fact the readers of this book are probably wondering why we are using wind energy rather than solar energy to produce hydrogen fuel.

First of all, there are many forms of solar energy, and wind is one form. Since without the sun to create temperature differences on the earth, there would be no wind. The sun's energy can also be stored, as you probably know, from the heat stored in the walls of your house. This stored heat causes the inside of your house to be warm at night when the outside temperature has dropped. Another form of stored energy from the sun is ocean thermal gradients. In this case, the surface temperature of the ocean water can be much warmer than the deep-subsurface temperature, and energy can be created by this temperature difference—as described in subchapter 3.7.

But let's get back to the question of why not use sunlight rather than wind as our chosen primary source of energy to produce hydrogen. To begin answering this interesting and logical question, we first need to determine how much of the sun's energy can be obtained for a given amount of land area.

Just as with the wind map shown in subchapter 3.2, measurements of the sun's energy have also been made for different parts of the world. These measurements are expressed as "insolation" values. For instance, when you look at the wind map, you will notice that the southeastern portion of the United States does not have very high sustainable winds; however, the "insolation" values in the southeast are high relative to the high-wind areas—like North Dakota. For comparison, the "insolation" value for Miami, Florida—where the wind level is low—has a yearly average of about 5.50 kilowatt-hours per square meter per day where in western North Dakota—where the winds are high—the average yearly "insolation" value is about 3.68 kilowatt-hours per square meter per day.

"Insolation" values are the measured energy from the sun over a period of time. For our evaluation, we will assume the average energy for a year since this best determines the amount of energy that can be expected on a repeatable basis for a given area. Therefore, let's begin our wind-versus-sun comparison in western North Dakota. Since, in our subchapter 3.1 analysis, we plan to obtain 15,625 kilowatts of average energy per square kilometer of land area, this will be our reference point.

To determine the maximum sun energy in a square kilometer, we need to multiply the North Dakota "insolation" value (3.68) by 1,012,082 square meters per square kilometer that equals 3,724,460 kilowatts-hours per square kilometer per day. Now we need to divide by 24 hours per day to make our comparison, which is 155,186 kilowatts of average energy per square kilometer of land area. At first glance, it looks like solar energy wins. But wait a minute, we now need to determine how much of the available sun energy we can collect and convert into electricity to make hydrogen. So let's discuss efficiencies.

There are numerous ways of collecting the sun's energy and transforming it into electric energy. One way is for mirrors to focus the sun's energy to create steam that generates electric power like a conventional power plant. The best way, however, appears to be by using photovoltaics—the direct conversion of sunlight into electricity. This is the method used by NASA to power everything from the Hubble telescope to the Mars rover.

While sun-tracking mirrors can be employed, the cost and simplicity of photovoltaics appears to be the favored approach among those who have evaluated each method—including profit-making companies who are growing by about 20 percent per year. In fact, the most favored approach used by these companies is the use of what is called "thin film" photovoltaics. While I have not made the economic evaluations of each alternative, I will assume that the future is in the form of "thin film" photovoltaics.

Current "thin film" photovoltaics range in efficiency from between 5 percent on the low end to 12 percent on the high end. However, 30 percent and higher efficiencies have been achieved with multijunction research cells. These research cells use exotic materials such as gallium arsenide or indium selenide and are predicted, in low volume, to cost about one hundred times more than mass-produced 8 percent efficient amorphous silicon cells with an output of only four times the electric power. Again, I am not in a position to dispute the cost comparison, so we will assume mass-produced "thin film" photovoltaics.

Even though "thin film" photovoltaic efficiencies currently vary from 5 percent to 12 percent, I will be optimistic and assume the Department of Energy (DOE) research goals of 15 percent efficiency and $50 per square meter. Therefore, based upon a 15 percent efficiency, the maximum power output per square kilometer in western North Dakota is 23,280 kilowatts of average energy per square kilometer of land area (155,186 x 0.15 = 23,280). At this point, our solar energy still looks OK when compared to our 15,625 kilowatts of wind energy.

But wait a minute, there is another glitch. First of all, we cannot use 100 percent of the available land area, and second, we need to further derate the output energy by about 20 percent to account for a dilution of the "insolation" value.

Let's assume that we can use 60 percent of the available land area because of installation, service access, and other factors. Then because the measured "insolation" value assumes that the collector is facing directly toward the sun and includes times where the sun is rising over the horizon or sinking below the horizon, a derating factor needs to be applied. An industry-accepted derating value of 20 percent is used that also includes such factors as dust interference.

Now let's multiply the 23,280 kilowatt value by 0.60 to account for land usage and by 0.80 to account for the "insolation" value derating. The result is 11,200 kilowatts of average energy per square kilometer of land area—in western North Dakota. At $50 per square meter—the Department of Energy (DOE) objective—the expected cost is $30.3 million ($50 x 1,012,082 square meters per square kilometer x 0.60 land use = $30.3 million). In comparison, the mass-produced 5-megawatt Magenn air rotor in western North Dakota achieves 15,620 kilowatts of average energy per square kilometer at a cost of $31.2 million.

Repeating the same calculations for the Miami, Florida, area at an average yearly "insolation" value of 5.50 kilowatt-hours per square meter per day, the result is 16,700 kilowatts per square kilometer for the same cost of $30.3 million—a better result than for our wind power; but remember, current thin film technology has not yet achieved a 15 percent efficiency.

In any event, as previously stated, in the subchapter 3.1 analysis, we will supplement the wind-activated electric generators with a limited quantity of "thin film" photovoltaics because of their "buffering" capability. An arbitrary number at this point is to purchase $100 billion of photovoltaics at $50 per square meter. This results in one billion square meters of photovoltaics. As a result of this investment, we will help provide make-up capacity for long durations of low wind conditions and stimulate the technology and economics of solar power. To evaluate the possibility of a much lower future cost, let's first look at the three competing technologies being pursued in the development of photovoltaics:

1. Amorphous silicon—being developed by Solarex, United Solar, Cannon, and others.
2. Cadmium telluride—being developed by First Solar, BP Solar, and Matsushita.
3. Copper indium diselenide—being developed by Siemens Solar Industries (SSI) and Global Solar.

Although these materials are not limited by world resources, they are relatively expensive to produce and generally comprise about 50 percent of current photovoltaic cost. If purchased in large quantities, the prices should come down dramatically.

The next cost issue is that in order to manufacture the "thin film" material, a process involving many layers of materials is required:

1. A transparent front layer that protects from the environment
2. A transparent and conductive top layer, or grid, that carries current away
3. A thin (1-4 micron) central sandwich of semiconductors that form one or more junctions to separate charge. This sandwich alternates n-type and p-type layers to form a junction
4. A back contact that is often a metal film
5. A back sheet that protects from the environment and that could be supportive (rigid or flexible)
6. Various intermediate processing steps: scribes and depositions to interconnect strip cells, annealing steps to activate or complete certain components, lamination to attach encapsulation, buss bar attachment to carry off power, isolation scribes at the borders, glass and other substrate handling, cleaning, and heating

These layering steps differ for each of the aforementioned photovoltaic types and require different time periods for drying, cleaning, curing, and heat treatments that are generally not conducive to increased production speed.

Whether or not the economic production of photovoltaics is feasible is a subject of evaluation during the project implementation stage. However, with a production requirement of 22,830 ten square meter panels per day for twelve years, we will begin to approach the volumes necessary to test the advanced manufacturing method in reducing costs.

During the photovoltaic feasibility study, we need to evaluate another possible solar hydrogen producing technology, which involves focusing sunlight as a method of splitting water into hydrogen and oxygen. Included in this study should be an evaluation of the use of titanium dioxide as a catalyst to enhance the water-splitting process. No more will be said about this, but it is mentioned here because it may provide a more cost-effective solar solution for producing hydrogen

Whatever the outcome of these feasibility studies, we can proceed with or without photovoltaics or other forms of solar energy.

Electrolysis

Typical Water to Hydrogen Electrolysis System

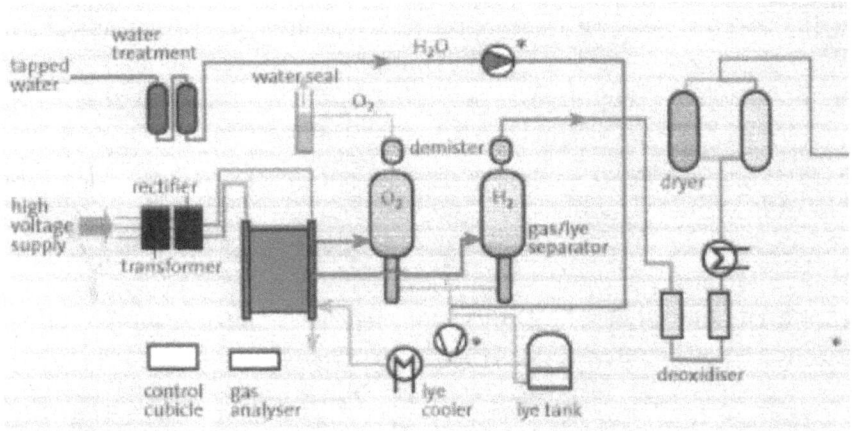

Hydrogen electrolysis is the process of passing an electric current through water to split individual water molecules into their constituents—hydrogen and oxygen. Energy efficiency can range from 65 percent to 85 percent depending on the type of equipment used. This means that as much as 35 percent of the electrical energy that is used to generate hydrogen and oxygen is lost in the process. In our subchapter 3.1 analysis, we have assumed that it will take 55 kilowatt-hours of electric energy to produce one kilogram of hydrogen gas—pressurized to 30 bar (435 psi) and 99.995 percent purity (a condition needed for liquefaction, since one kilogram of hydrogen gas contains about 36 kilowatt-hours of energy; the assumed process efficiency equals 65 percent (36 / 55 x 100 = 65 percent).

There are three currently available variants for electrolytically producing hydrogen from water. These are the following:

- Alkaline water electrolysis with aqueous potash lye as the electrolyte. The operating temperature is about 80 °C (176 °F).
- Membrane electrolysis with a proton-conducting membrane as the electrolyte. The basic design is similar to a fuel cell, and the operating temperature is about 80 °C (176 °F).
- High-temperature steam electrolysis that uses oxygen-ion conductors as the electrolyte. The operating temperature is between 800 °C (1472 °F) to 1000 °C (1832 °F).

Alkaline Electrolysis

Alkaline electrolysis is currently used to provide hydrogen for chemical industry processes—metallurgy, fat hardening, and electronics—in regions where electricity is available at favorable prices

The capacity of this equipment has reached more than 33,000 Nm3/hr (1,254,000 standard cubic feet per hour) at 30 bar pressure (435 psi) and 65 percent to 70 percent efficiency. This equals 6,530 kilograms of hydrogen per hour, and our requirement is 417 kilograms per hour at 65 percent efficiency. Therefore, this equipment—scaled down and adapted to produce 99.995 percent purity—would suit our needs.

In the 1980s, a number of research programs were conducted with the aim of increasing the efficiency. It was found that plasma electrode activation leads to a significant efficiency increase. More than 80 percent efficiencies were achieved, and industrial applications have proven lifetimes of more than 40,000 hours (4.57 years). Also, within the framework of what is called the HYSOLAR project, 85 percent efficiencies have been achieved from fluctuating energy—like our wind-generated electricity will be.

Membrane Electrolysis

Membrane electrolysis is used primarily to make small quantities of on-site hydrogen for use as a fuel. A commercial "membrane type" electrolysis product "HySTAT"—produced by the Hydrogenics Corporation—is designed to make a kilogram of 99.995 percent pure hydrogen gas at 30 bar pressure or higher using 53.22 kilowatt-hours of electric energy. This is 1.78 kilowatt-hours less than our specified 55 kilowatt-hours, or slightly better than 65 percent efficient.

The largest standard HySTAT product produces 11.9 kilograms per hour of hydrogen, and we need 417 kilograms per hour—as specified in our subchapter 3.1 analysis—which means that we need to increase the capacity of the HySTAT product by about thirty-five times. My discussions with a Hydrogenics representative indicate that efforts are underway to increase their product capacity to meet the 417 kilograms per hour goal. As a result, the mass-produced price for this product would be very much in line with the numbers used in my subchapter 3.1 analysis.

High-Temperature Steam Electrolysis

High-temperature steam electrolysis is currently in the development stage. The process operates at between 800 °C (1,472 °F) and 1000 °C (1,832 °F) and uses a natural gas-assisted steam electrolyser that uses the partial oxidation of natural gas to synthesis gas on the anode side. In this case, no oxygen evolves on the anode side, but oxygen is directly used for the partial oxidation of hydrocarbons. The low electrochemical potential of this "natural gas-consuming anode" permits very low cell voltages for steam electrolysis from a thermodynamic perspective and thus a high efficiency.

Because this process uses natural gas—a limited resource hydrocarbon—and it is currently in the development stage, I have not selected high-temperature steam electrolysis for our project.

Conclusion

Since the alkaline electrolysis method is commercially available in the capacity that we need, I have selected this method; however, if the 417 kilograms per hour HySTAT product becomes available at the projected pricing, this would be an alternative—especially at a slightly lower energy consumption rate. However, to avoid controversy and to account for water pumping and hydrogen purification requirements, I will not assume an efficiency greater than 65 percent. If an efficiency of 80 percent or more is deemed possible during the feasibility study portion of the program, we can make the adjustment. At 80 percent efficiency, the amount of electricity needed to produce 1 kilogram of hydrogen gas would be 45 kilowatt-hours versus 55 kilowatt-hours used in our analysis.

Again, to evaluate the cost of the electrolysis equipment, I have estimated the cost of a 417 kilograms per hour of pressurized and purified hydrogen gas machine at $10 million. This is equivalent to purchasing 400 automobiles for $25,000 each.

Superconductivity

Combined Hydrogen Pipeline—Transports Gaseous Hydrogen, Liquid Hydrogen, and Electricity

In our reference system, described in subchapter 3.1, I proposed that we use superconductivity to deliver the electricity needed to make liquefied hydrogen at the end of the hydrogen gas pipeline. Superconductivity is

a method of transmitting electricity at very cold temperatures. The cold temperature allows electrons to pass through wires much more easily than they would in standard ambient conditions. So much so that, as stated before, in one instance, 18,000 pounds of copper wire cable was replaced by 250 pounds of superconducting wire to carry the same power. In addition, about five times as much electricity can be transmitted per superconducting wire cable, and the transmission losses can be as much as twenty-five times less. In our subchapter 3.1 analysis, I assumed that as much as 15,000 megawatts would be transmitted per superconducting line.

There are two superconducting methods that could be considered for this project. The first uses liquid-nitrogen cooling in a closed-loop pumping system (a 30-kilometer version is successfully operating in Denmark today). The second method uses liquid hydrogen in an open-flow system with gravity-assisted pumping and hydrogen reprocessing stations along its path. There are advantages and disadvantages to each approach that need to be evaluated, but the choice that appears to make the most sense is the open hydrogen-cooled system. This is because liquefied hydrogen can be economically produced at electric generating sites and utilized for fuel at the end of the transmission line.

The biggest disadvantage in transmitting liquefied hydrogen is the "boil off" losses, which can reach 3 percent or more per day—especially as it creates frictional heat along its path. And when you consider how much volume of liquefied hydrogen could be contained in a 3,000-kilometer pipe, the "boil off" losses could be very significant. Therefore, to make the system viable, it would be necessary to capture the "boil off" and reprocess it. Also, because we are not restricted to the amount of liquefied hydrogen that is transmitted, we can reduce frictional losses by operating at very low flow rates. In any event, an economic trade-off would be necessary to get the best "bang for the buck."

Not being familiar with all of the design aspects of the open hydrogen superconducting system, I recommend that this aspect be the purview of experts. Obviously, the number and cost of "boil off" reprocessing and pumping stations would be a factor—including rerouting and maintenance requirements. In any event, the cost should be less than overhead electric transmission lines using high-voltage direct current (HVDC)—commonly used to transmit bulk electric power. For the purpose of establishing the potential cost of the system, a planned HVDC transmission line was used as a reference. The planned system was a 4,000-kilometer HVDC line from a Sahara wind farm to Europe—that included an underwater connection between Africa and Europe.

Hydrogen Gas Transportation

In this section, we are going to look at bulk transportation of hydrogen gas via pipeline (our reference design) and by zeppelin technology.

Pipelines

Hydrogen liquid or gas can be transported via pipeline. Since liquefied hydrogen could be transported via pipeline, it suffers from losses due to "boil off," pumping, and friction (if carried at high flow rates). These losses make pipeline transport of liquefied hydrogen somewhat impractical. However, as discussed in the previous section, we will transport some liquefied hydrogen because of its superconductivity benefit. In this transport mode, an economic balance between flow rate and "boil off" will determine how much liquefied hydrogen is transported and not the usage requirement.

Hydrogen gas can be transported by pipeline in a manner similar to how natural gas is transported today. However, natural gas pipelines are generally not suited for conversion to hydrogen gas. Three reasons for this are as follows:

1) Natural gas pipelines are generally made of low-carbon steel, which would be subject to hydrogen embrittlement and subsequent cracking when used with hydrogen—aluminum, stainless steel, alloy carbon steel, or aluminum wrapped in a composite material would probably be the material of choice for a hydrogen pipeline.
2) Because of its low density, hydrogen gas requires about 3.8 times more pumping energy to propel the same volume of hydrogen gas through the same-size natural gas pipelines; therefore, larger pipelines would be needed to reduce the friction losses and pumping energy. However, larger more expensive screw compressors would be required.

3) Hydrogen gas pipelines and fittings need to be welded and sealed in a more stringent and controlled manner than has generally been used with natural gas pipelines.

Because hydrogen's density is about one-third that of natural gas, it would be necessary to make the hydrogen gas pipes about 65 percent larger in diameter (because of the diameter-squared effect) to transmit the same quantity of hydrogen gas as natural gas. However, the frictional losses for hydrogen would be reduced by about 60 percent—when using the 65 percent larger diameter pipe—because of its reduced ratio of contact (friction) surface area to flow volume. Therefore, because pumping energy is primarily a function of frictional losses, the required pumping energy would be less for the hydrogen gas. Furthermore, because the piping thickness increases in proportion to the increased diameter, an economic trade-off would be needed to determine the optimum piping diameter.

Given this background, we can conservatively say that hydrogen gas can be transported over long distances with the same, or less, energy losses as natural gas, but at higher piping, compressor, valving, and fitting cost—because of their larger size.

Studies show that hydrogen gas can be transported at 68 bar (1,000 psi) in 915 mm (36 in) and 305 mm (12 in) pipelines, with flow rates of 100,000 kilograms of hydrogen per hour and 10,000 kilograms of hydrogen per hour, respectively. Extrapolating these numbers to larger pipe systems, we can expect to transport about 400,000 kilograms per hour in a 1,830-mm-diameter (6-ft) pipe.

Since we need to deliver 1,970,000 kilograms of hydrogen gas per hour, in our subchapter 3.1 analysis, it would require five 1,830 mm pipelines (with smaller diameter branches). One of the side benefits of so many large diameter pipes is that the stored hydrogen "buffer" energy in a 3,000-kilometer hydrogen gas-pipeline system is about 4.4 days of hydrogen consumption—a great benefit to having a large-diameter pipe system because of the difficulty in storing hydrogen. The stored energy is the amount of hydrogen that could be consumed by letting the 68 bar pressure drop to its minimum transport pressure level.

To determine the cost of the 3,000-kilometer five-pipe system, we will first calculate the material cost of five 1,830-mm-diameter, 25.4-mm-thick (1.0-in) alloy carbon steel pipes. If we could produce the required alloy carbon steel piping at $0.70 per kilogram, the material's cost alone for the five pipes would be about $12 billion.

To determine the installation cost, I will use the Alaska pipeline as a comparative example. The Alaska pipeline is 1,280 kilometers (800 miles) long. It is 1,220 mm (48 inches) in diameter and crosses over eight hundred rivers and streams and a number of mountain ranges. Construction started in March 1975 and had 720 kilometers (450 miles) completed by the end of 1976—less than two years. By May 1977, all 1,280 kilometers were completed. Employment reached a peak of 21,000 in August 1975, and the total cost was $8 billion (1977 dollars). Using 21,000 workers at $50 per hour for 10 hours per day and a three-year time frame as a reference for the hydrogen gas pipeline, we would need about $11.5 billion for labor cost. Increasing this amount, for the requirement to have five larger diameter pipes installed, we will assume $15 billion for labor cost.

When adding $12 billion for materials and $15 billion for labor, the result is $27 billion. Since our subchapter 3.1 analysis assumes $40 billion, we have $13 billion remaining to pay for the superconducting transmission line, hydrogen gas pumps, and other expenses.

Zeppelins

Because hydrogen is light in weight, it is uniquely capable of being transported by air, using zeppelins (refer to subchapter 4.2). To understand this better, consider the fact that 1 kilogram of hydrogen is about equal to one gallon of gasoline, and the gallon of gasoline weighs 3.05 kilograms—a three-to-one weight ratio. However, even in a liquefied form, hydrogen occupies four times as much volume as gasoline of equal energy content, thus making hydrogen more difficult to transport by truck, train, or even by ship than gasoline. With a zeppelin, it is only the weight—not volume—that matters.

Although liquefied hydrogen weighs less than gasoline, a hydrogen tank will weigh more than a gasoline tank made from the same materials. The amount of additional weight for the liquefied hydrogen tank is very dependent on the contained volume and the materials used for construction.

For small volumes, the added tank weight—using aluminum—could exceed the hydrogen weight by more than ten—this was discussed in subchapter 3.5 relative to vehicle fuel tanks. However, for very large volumes, the tank weight contribution could be equal to or less than the contained liquefied hydrogen weight. Therefore, since zeppelins will carry very large volumes, we will assume a full weight of twice the contained hydrogen. The result would then be that transportation of liquefied hydrogen by zeppelin would be about 60 percent of the equivalent energy weight of gasoline—when you factor in the weight of the gasoline container.

Here is where transportation by zeppelin starts to make sense. Why? Because a zeppelin can be made very large, and its lift capabilities can increase very rapidly with increasing lengths and diameters.

In subchapter 3.1, it was determined that we need to deliver 47.3 million kilograms (104.0 million pounds) of liquefied hydrogen per day from each of ten sites. Since we need to double the hydrogen weight to account for the tank weight, the weight of liquefied hydrogen-filled tanks would be about 94.6 million kilograms (208 million pounds). Using the 1.5-million-pound carrying capacity of the zeppelins described in subchapters 4.1 and 4.2, it would take about 280 zeppelins making one round-trip in two days. At $10 million per semi-self-powered (with photovoltaics) zeppelin, the capital cost would be almost $3 billion. When compared with the hydrogen gas pipelines at $27 billion, it is clear that zeppelin delivery is much less expensive—even if hydrogen pumping versus zeppelin fuel costs are taken into account. Perhaps some of my proposed pipelines should be replaced with zeppelins!

Liquefaction

Hydrogen gas has to be cooled to about 235 °C (−391 °F) to transform it from a gas to a liquid. This requires a multistage refrigeration process that consumes electricity without increasing the energy content of the gas. This increased energy is assumed to be 15 kilowatt-hours to liquefy 1 kilogram of hydrogen. However, as shown in the graph below, 13 kilowatt-hours is not out of reach for our required capacity of 417 kilograms of hydrogen per hour.

To understand the graph, let's look at 417 kilograms per hour (kg/h) and find the ratio of liquefaction energy to the higher heating value (HHV) of hydrogen. This ratio is approximately equal to 30 percent for the best liquefaction methods. Since the energy content of hydrogen (HHV) is about 36 kilowatts-hours per kilogram, then the liquefaction energy is 10.8 kilowatts-hours (36 x 0.30 = 10.8). In other words, the ratio of 10.8 kilowatt-hours to hydrogen energy content of 36 kilowatt-hours is equal to 30 percent—(10.8 kilowatt-hours of electricity / 36 kilowatt-hours of hydrogen energy content = 30 percent). Therefore, a conservative estimate of 15 kilowatt-hours per kilogram to convert hydrogen gas into liquefied hydrogen gas was selected to avoid controversy.

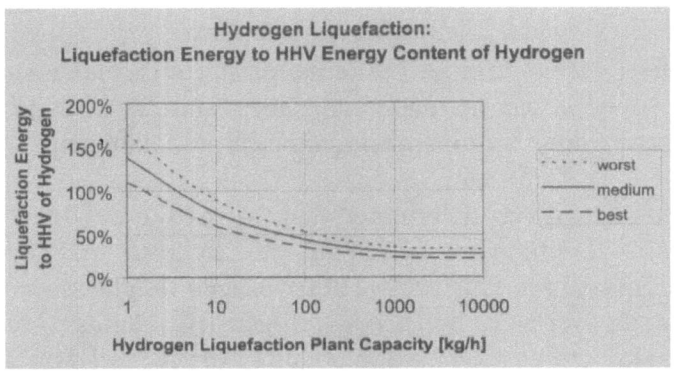

The reason for liquefying the hydrogen gas is to overcome one of hydrogen's biggest problems—volume. Liquefied hydrogen can be used in a manner similar to gasoline, but it still has a volume four times as great as an energy-equivalent volume of gasoline. This fact in itself is interesting in that gasoline contains hydrogen, and it is mostly the hydrogen that burns as a fuel and not the carbon—the carbon leaves the combustion process, mostly, as carbon monoxide (CO) and carbon dioxide (CO_2). Therefore, gasoline appears to be one of the best ways to contain and store hydrogen. This leads to my belief that there may be a man-made product in the future that will store hydrogen on a molecular level such that liquefaction of hydrogen will not be necessary. That man-made product may be carbon *nanotubes*, and as previously mentioned in subchapter 3.5, research is underway in numerous laboratories to make it an economic reality. However, for the present, let's use liquefied hydrogen and hope for another solution in the future.

In simple terms, the liquefaction process begins by transforming air into liquefied nitrogen through several steps of compression, cooling, and expansion. Using the liquefied nitrogen, the hydrogen gas is precooled in several steps to about 193 °C (−315 °F) to what is called the Joule-

Thomson effect. Further cooling takes place through several compression/ refrigeration steps that results in part of the gas stream becoming liquid. The remaining gas is used in the upstream refrigeration processes as a coolant in the condensing part of the refrigeration cycles. During the final stages of cooling, an ortho-para conversion is performed by catalysts at four different temperature levels.

Industrially used liquefaction systems are based upon what is called the Claude process (see the Claude system flow diagram below). The liquefaction capacity is approximately 4,300 kilograms of hydrogen per day at a specific power level of 13.4 kilowatt-hours per kilogram of hydrogen—not too different from our 10,000 kilogram per day and 15.0 kilowatt-hour per kilogram requirement.

Development is currently underway to improve efficiency to a specific power of 5.0 kilowatt-hours per kilogram by using a magnetocaloric refrigeration processes and shifting the ortho-para conversion to higher temperature levels with the aid of electromagnetic catalysts. However, the conventional Claude process is available now and is therefore our selected method—this again avoids controversy.

Claude system flow diagram

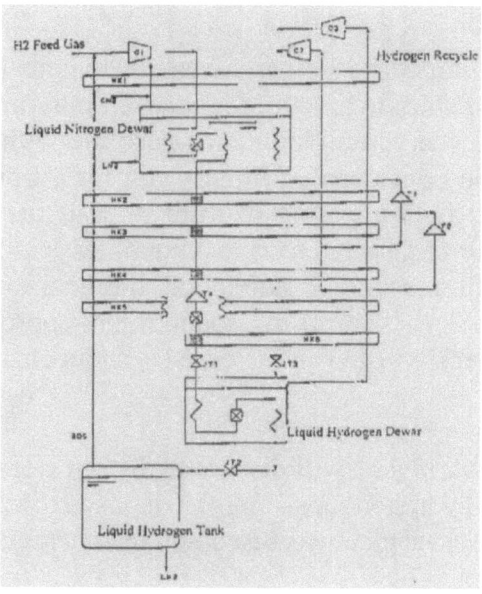

Please excuse the technical discussion, but it is provided here with the intent of giving the reader an idea of some of the complexity involved in the liquefaction process. As such, it points out the requirement for a number of compressors (electric-motor driven), expansion valves or turbines, and condenser/evaporator heat exchangers, liquid separators, mixing tanks, vacuum devices, and holding tanks, which in turn helps describe the nature of the machine and some idea of potential cost.

Based upon this information, and my own experience with the cost of refrigeration equipment, the flow rate requirements of the various components, and *advanced* mass production techniques, it is not unrealistic to assume the price of a 10,000-kilogram per day (417 kilogram per hour) machine to be less than $2 million each—installed. In fact, an estimated cost of less than $400,000 is considered to be conservative. In addition, to continue with the automobile visualization of cost, $2 million is equivalent to purchasing 80 automobiles at $25,000 each.

Storage

Hydrogen is the Houdini of all substances—it wants to escape from wherever it is contained. It is also the least dense of all substances that makes its volume very large relative to its weight. This makes for a formidable challenge, to say the least.

As previously mentioned, it is rather interesting to note that gasoline is one of the best ways to store hydrogen. Since, even in liquid form, hydrogen occupies four times as much volume as gasoline with the same energy content. When you consider that liquids are, for all practical purposes, incompressible, it is fascinating to think that at the molecular level, hydrogen's volume can be reduced to four times less than its liquefied form. Therefore, from a storage standpoint, our goal would be to confine hydrogen at the molecular level in a manner that approximates gasoline, but allows it to be easily filled and released without changing its chemical form.

In this section on storage, I will discuss hydrogen storage as it relates to "buffering" of the hydrogen fuel supply as opposed to storage of hydrogen in a vehicle fuel tank, which was discussed in subchapter 3.5.

Buffering

Buffering of the hydrogen fuel supply is required because the wind—and solar-generated electricity—used to make hydrogen—is not always available. In spite of all efforts to place the wind energy devices in locations of high sustainable winds, there will be times when the wind is not blowing hard enough to provide the amount of hydrogen fuel being consumed. As a first line of defense against a loss of wind energy, we have chosen to include supplemental photovoltaic solar panels to provide a daily supply of energy even when the wind energy is low.

Supplemental solar energy will still not be enough for long periods of low wind energy, even though the solar contribution will be reliable on a daily basis.

However, we do know from many years of monitoring the winds that the amount of wind energy can be relied upon in the long run. The problem is that when the wind does, or does not blow, the periods of feast or famine may last for weeks.

So what we could have is a lot of wind for weeks at a time and too little wind for weeks at a time. Thus the requirement for "buffering" the wind energy by storage of hydrogen when the wind energy is low and vice versa; when the wind energy is high, we need to replenish our storage of hydrogen. The amount of stored hydrogen can be determined analytically by knowing the wind and solar history for a given area and then multiplying that calculated amount by an estimated number to provide for contingencies.

If the calculated storage requirement shows that for some periods of time the supplied wind energy exceeds the required need for replenishment, the generated electricity will then be lost, unless it can be transformed and stored as hydrogen or supplied to the national grid system to save on electric power plant fuel. Since our proposed hydrogen fuel infrastructure provides about 62 percent of our total national energy consumption needs (intended for vehicle use and not for generation of electric power or other uses), our national electric supply system should be able to absorb excess electric power—beyond that needed for replenishment of stored hydrogen.

For the case where too little electric power is produced to meet the amount required for fuel consumption, we will buy back power from the national electric grid system. This is generally possible since electric power plants are capable of meeting peak electric demand, which means that they have about twice as much electric-generating capability as is needed for average consumption needs, and we can use this power during off peak times.

Based upon these contingencies—supplemental solar and national electric grid supply—it would seem at first glance that a "buffer" storage system would not be necessary. Well, let's look at the numbers. What if we could only supply 60 percent of the hydrogen fuel demand for one month of low wind energy? This means that we would need another 40 percent of electric power from the national grid if there were no buffer storage.

Let's do the arithmetic. Our current national average daily hydrogen fuel consumption is assumed to be 471 million kilograms of hydrogen per day. With 60 percent being supplied, we will have a shortfall of 189 million kilograms per day. The 189 million kilograms of liquefied hydrogen requires 550 million kilowatts of electric power generation capability based upon 70 kilowatts-hours per kilogram of liquefied hydrogen. Since the entire U.S. national electric power supply is about 880 million kilowatts of electric-generating capacity (i.e., twice the average usage), it will not be possible to make up all of the required electric power needs (i.e., half of 880 million kilowatts—if available—is 440 million kilowatts, and 550 million would be needed—a shortfall of 110 million kilowatts or 37.6 million kilograms of liquefied hydrogen shortfall per day).

However, the wind-generating areas would be distributed and interconnected, such that it can be expected that where one wind area is not producing, another one will. With this consideration, let's assume that we need 40 million kilograms of liquefied hydrogen makeup per day for one month of shortfall.

It will be shown in the following section that we will have five days' supply of liquefied hydrogen at each central station and one day's supply of liquefied hydrogen at each gas station. In addition, the hydrogen gas pipeline will have an additional 4.4 days' supply of gaseous hydrogen—for a total of 10.4 days' supply of stored hydrogen.

If one day's supply equals 471 million kilograms of hydrogen, then our total of 10.4 days' supply equals 4.9 billion kilograms. When we divide 4.9 billion kilograms by our shortfall of 40 million kilograms needed per day, we get a "buffer" capability of 123 days—assuming that the buffer tanks are full, and the hydrogen pipeline is at full pressure. Since our first priority will be to keep the pipeline at full pressure, we can expect that—on average—the liquefied hydrogen tank storage will be about half full, about three days', rather than six days', supply. Therefore, the "buffer" storage at the time of the shortfall will instead be about eighty-seven days.

Conversely, when the wind-generating areas are providing a maximum power output, it will not be possible for the electric grid system to absorb all of the excess power—as much as 687 million kilowatts of excess power output, thus requiring an additional method of storage. That's where the six days' supply of liquefied hydrogen tanks and interconnection of wind-generating areas come into play.

From the above analysis, it becomes clear that the national grid system can play a significant role in mitigating excess electrical energy supply shortfalls, but it is not going to be the entire solution, thereby making it necessary to be either refilling the depleted storage tanks or supplementing underperforming wind energy areas.

This is of course an idealistic scenario, and only experience will be able to determine the correct proportion of "buffer" capacity. In any event, as a matter of national security, it will be necessary to have a strategic national reserve.

Strategic Reserve

Currently the United States has a strategic petroleum reserve of about thirty-seven days. This is based upon the known domestic reserves of oil that can be produced in an emergency. For a country like Japan, which has no oil reserves, the strategic reserve is 171 days. In our hydrogen fuel economy—like Japan—we will have no back up reserves of hydrogen fuel. To compensate for this, the United States will have plenty of hydrocarbon reserves that can be converted to hydrogen when needed.

Therefore, the United States oil reserve of thirty-seven days is probably sufficient. However, this reserve that is located in four salt mines along the gulf coast of the United States needs to now have the capability of reforming oil into hydrogen and distributing it via pipeline (or other alternative) to the locations where it would be needed.

As stated earlier, gasoline, or oil, is one of the most compact ways of storing hydrogen. It almost becomes impractical to consider long-term storage of hydrogen in a gaseous or liquefied form. Gaseous hydrogen because it consumes too much volume and requires pressurization, and liquefied hydrogen because it also contains too much volume and because it must be used within a few days because of "boil off" considerations.

Reforming oil or gasoline to form hydrogen requires steam and a process temperature of 700 °C to 1000 °C (1292 °F to 1832 °F). The chemical reaction for gasoline is as follows:

$$C8H_{18} + 8H_2O \text{ (+ heat)} \rightarrow 8CO + 17H_2$$

Other hydrocarbons form hydrogen in a similar manner, so the process is relatively simple and straightforward. For instance, natural gas is the most common way of producing hydrogen since its ratio of hydrogen atoms to carbon atoms is large relative to other hydrocarbons. The chemical reaction for natural gas is as follows:

$$CH_4 + H_2O \text{ (+heat)} \rightarrow CO + 3H_2$$

The problem that arises is the investment needed for equipment and pipelines to produce and deliver the hydrogen to where it is needed.

If a technological breakthrough should occur in the future—like carbon nanotubes or sodium borohydride—it may be desirable to revisit the decision to strategically store hydrogen as oil or even using liquefied hydrogen in the hydrogen economy. But the nanotube—or other storage medium—requirement to duplicate oil as a low-pressure, low-cost storage medium is far from being a reality. In fact, the 2015 target goal of the Department of Energy (DOE) for small-size vehicle fuel tanks is an energy storage of 2.7 kilowatt-hours per liter of tank storage volume and

3.0 kilowatt-hours per kilogram of filled tank storage. In comparison, oil storage energy (in an equivalent small storage tank) is about 5 kilowatt-hours per liter of storage tank volume and 9.5 kilowatt-hours per kilogram of filled storage tank.

Therefore, for the strategic storage of hydrogen, the current thirty-seven-day oil storage with backup of inground domestic oil reserves is the most prudent and least disruptive method compared to anything in the foreseeable future. This strategic reserve would need to be supplemented by reforming equipment and distribution pipelines. And since we are using hydrogen gas pipelines anyway—in the hydrogen economy scenario—the strategic pipeline tie-ins should not be an overwhelming obstacle.

Delivery

In this section, we will examine how the gas station receives and dispenses liquefied hydrogen.

As mentioned earlier, the gas station receives liquefied hydrogen from a central processing and dispensing station. For simplicity, we will assume that a central dispensing operation services about fifty gas stations within a radius of about 160 kilometers (100 miles). For further simplicity, we will assume that each gas station dispenses about 5,000 kilograms of liquefied hydrogen per day (remember—one kilogram of liquefied hydrogen is equal to about one gallon of gasoline).

The central processing and dispensing station is owned and operated by an oil company. Therefore, there may be more than one central station in a given area. This may not be the most cost-effective way to do it, but we need to have free market forces to assure efficient and effective performance.

Each central station will receive low-cost electric power and pressurized hydrogen gas from "branches" of the pipeline and then convert the hydrogen gas into liquefied hydrogen, using low-cost liquefaction equipment. Keep in mind that this equipment was deliberately selected to produce liquefied hydrogen at the rate required by the average gas station, so the option exists for the gas stations to produce hydrogen fuel on site.

The central station will store five days' supply of liquefied hydrogen to use as a buffer against power reductions caused by low wind conditions. Each tank will be equipped with the Linde "CooLH2" system to prevent "boil off" for twelve days.

The central station will own and operate a fleet of specialized double tractor-trailers that are designed to carry 5000 kilograms[*] of liquefied hydrogen per trip.

Each 5,000-kilogram-per-day gas station will have six 2,500-kilogram capacity tanks. This will provide a one-day buffer supply for contingencies. In other words, two full tanks are delivered, and two empty tanks are returned each day—thus leaving two tanks in reserve. Note: Since the tanks are positioned vertically, it may be more practical to use two shorter 1,250-kilogram tanks per trailer.

To make the drop-off and pickup as fast and as least disruptive as possible, the reserve tanks should be connected and the empty tanks disconnected prior to arrival of the full tanks. Robotic equipment should be utilized in this process.

The number of double tractor-trailers owned and operated by the central station would depend on demand requirements. Since, in our example, each central station services fifty gas stations within a 160-kilometer (100-mile) radius, a maximum round-trip time per vehicle would be about

---

[*] A standard gasoline truck can deliver gasoline with the energy equivalent of 8,537 kilograms of liquefied hydrogen per trip. Since liquefied hydrogen provides 25 percent to 100 percent or more fuel efficiency than gasoline, a 5,000-kilogram delivery of liquefied hydrogen would be equivalent to about 6,250 to 10,000 gallons of gasoline. Each trailer would be equipped with a lifting device that can lift the liquefied hydrogen-filled tank to a vertical position. The empty tank would be then lifted, with the same device, back onto the trailer and returned to the central station for refilling. Since some gas stations are located in traffic-congested areas that will not permit the full-length double trailer system to make deliveries, it would be possible to make two deliveries by leaving the rear trailer at a nearby parking lot and picking it up on a second run—a common practice in Europe.

6 hours (at an average of 64 kilometers per hour—40 miles per hour—for a 320-kilometer [200-mile] round-trip would be 5 hours plus an hour for delivery). Therefore, each vehicle could make three trips per day if operated 24 hours per day. If the average double tractor-trailer services one gas station per trip, the maximum number of vehicles required per central station would be seventeen (50 / 3 = 17). Let's make it twenty to account for repairs and traffic delays.

There are many other scenarios for production and delivery of liquefied hydrogen fuels, including production and dispensing of pressurized hydrogen gas. But as you can see, because of the close proximity of the central stations to the gas stations, there could be a distinct advantage over gasoline. This is because when gasoline pipelines are not available, the gasoline needs to be delivered from the refineries to the gas station, so fuel delivery costs could be higher for gasoline. Furthermore, when you consider that oil has to be delivered to the oil refineries from the oil reservoirs—usually by means of huge oil tankers—it is not hard to imagine how domestically produced hydrogen gas delivered to central stations around the country could have a distinct advantage.

One question that you might have is, will the government, oil companies, and gas stations make as much money dispensing hydrogen versus dispensing gasoline? The answer to this question is not clear, since 25 percent to 100 percent less fuel will be consumed with hydrogen vehicles. Therefore, an adjustment may need to be made with regard to federal/state taxes and profit per kilogram of hydrogen sold. However, since the complete implementation of the hydrogen fuel infrastructure will not occur for sixteen to twenty years, and with a current U.S. consumption rate increase of 1 or 2 percent annually, a tax and profit adjustment may not be necessary. In addition, think of all the money that will be spent in the United States for hydrogen production rather than for imported oil. This alone should significantly increase the tax base.

One consideration for reducing oil and gas station cost is advertising; since hydrogen fuel will provide the cleanest burning high-test fuel, the application of distinguishing additives will probably not be possible. One distinguishing possibility may be a degree of purity that could be aimed at fuel cell vehicles—to make them last longer with less service problems. One other distinguishing idea may be improved gas station service—like it was in the old days—and faster fill times. In any event, it is my belief that most people consider gasoline to be a commodity that is distinguished only by a like or dislike of a particular gas station or oil company. Most of the time, it is price and/or convenience that will make the difference.

But what about the tremendous investments already made by the oil companies, in tankers, oil refineries, gasoline delivery trucks, and other gasoline-related equipment. Don't worry, all investments in capital equipment have a depreciation time period, and there will be plenty of notice given to the oil companies to phase out oil and gasoline-related investments within a twenty-year time frame. In fact, it may make sense to convert oil tankers into container ships or for exporting domestically produced oil to other countries. And if the *advanced* manufacturing idea catches on, we could see the United States becoming a net exporter of manufactured goods and an importer of goods for recycling, thus requiring more converted oil tankers.

My Background

My degree is in mechanical engineering from Northeastern University in Boston, Massachusetts. Upon graduation in 1964, I was commissioned a lieutenant in the United States Army Corps of Engineers and served in Vietnam.

In August 1966, I began my career with the Westinghouse Electric Corporation. During the first seventeen years, my experience was primarily in nuclear steam generator development. During the last four years of these seventeen years, I was promoted to Section Engineering Manager and, later, Department Engineering Manager of a $40-million Atomic Energy Commission (presently the Department of Energy [DOE]) contract to develop a steam generator for a "breeder" nuclear power plant. My department also performed government contract engineering studies to develop solar power steam generators and fluidized bed coal gasification boilers.

In 1982, I transferred to the Thermo King Corporation (a subsidiary of Westinghouse) to become Engineering Manager for the Truck Transport Refrigeration Equipment Department. During the early portion of my employment with Thermo King, my engineering group developed a new line of truck refrigeration equipment for the world market. I was later given responsibility for design engineering activities at Thermo King factories in Barcelona, Spain; Hamble, England; and Prague, Czech Republic (formerly Czechoslovakia). Before retirement in 2001, I was responsible for the successful development of a new line of refrigeration equipment for the Japanese market and the development of a new alternator-powered refrigeration system for the European market.

I have written numerous technical papers for international conferences and for the American Society of Mechanical Engineers (ASME) and currently hold seven patents. I was also the Chairman of the Florida West Coast Section of ASME in 1974-75 and passed the state of Florida professional engineering examination in 1975.

My experience in power generation equipment design, and the design and manufacturing of complex electromechanical equipment, makes me uniquely qualified to evaluate the process of producing hydrogen and the mass production costs of equipment associated with that process.

References

1. Lavelle, Marianne, The New Oil Rush, U.S. News and World Report, Money & Business, April 24 2006
2. Bossel, Ulf, Baldur Eliasson, and Gordon Taylor. *The Future of the Hydrogen Economy: Bright or Bleak?* Oberrohrdorf, Switzerland: Ulf Bossel 15 April 2003.
3. Keith, Geoffrey, and William Leighty. *Transmitting 4,000 MW of New Windpower from North Dakota to Chicago: New HVDC Electric Lines or Hydrogen Pipeline*. Draft report for Environmental Law and Policy Center, Chicago. Cambridge, MA: Synapse Energy Economics, 28 Sept. 2002.
4. National Research Council. Board on Energy and Environmental Systems. *The Hydrogen Economy: Opportunities, Costs, Barriers, and R&D Needs*. Washington, DC: National Academy of Sciences, 2004.
5. Harry W. Braun, *The Phoenix Project: Shifting from Oil to Hydrogen (www.phoenixproject.net),* Sustainable Partners International, Tel. 602-955-4555, Phoenix AZ, 2000.
6. Doty, David F. *A Realistic Look At Hydrogen Price Projections*. Doty Scientific Inc. 23 May 2004 *www.dotynmr.com/PDF/Doty_H2Price. pdf*
7. Kondoh, J., et al. *Electrical energy storage systems for energy networks*. Energy Conservation and Management 41.17 (2002).
8. Donalek, Peter. Advances in Pumped Storage. Presented at Electricity Storage Association Spring Meeting, Chicago, Il. 21 May 2003.
9. Vyas, Arant D., and Henry K. Ng. *Batteries for Electric Drive Vehicles: Evaluation of Future Characteristics and Costs through a Delphi Study*. Chicago, Il: Argonne National Laboratory, 1997.
10. DOE Hydrogen Program, *III.0 Hydrogen storage sub-Program Overview,* FY 2004 Progress report.

11. Hirsch, Robert L., Bezdek, Roger H., Wendling, Robert M., Peaking *Oil Production: Sooner Rather Than Later?*, Issues in Science and Technology, Spring 2005.

12. Boyle, Godfrey, *Renewable Energy*, Oxford University Press, 1996.

13. Deffeyes, Kenneth S., *Beyond Oil*, Hill and Wang, 2005.

14. Taylor D.Sc, Hugh, S., *Industrial Hydrogen*, Knowledge Publications, 2006

15. Romm, Joseph,J., *The Hype About Hydrogen*, Island Press, 2005.

16. Magenn Power Company, *Magenn Air Rotor System (MARS) Executive Summary*, April 2006.

17. HySTAT-A Hydrogen Plants, *Changing Power . . . Powering Change*, Hydrogenics Corporation Catalog, April 2006.

18. Tyndall Center, Hydrogen Economy Power Point Presentation, 2006.

19. Mazza, Patrick, Hammerschlag, Roel, *Carrying the Energy Future—Comparing Hydrogen and Electricity for Transmission, Storage and Transportation,* The Institute for Lifecycle Environmental Assessment, Seattle WA June 2004.

20. Scientific American Magazine—September 2006

21. Forbes Magazine—June 2006

22. Consumer Reports Magazine—October 2006

23. *http://en.wikipedia.org/wiki/Synfuel—9/6/2008*

24. *http:/dnr.louisiana.gov/sec/execdiv/techasmt/alternative_fuels/cng. htm 8/26/2008*

25. *http://www.crest.org/hydrogen/hydrogen_fuelcell_reformst . . . 4/5/2006.*

26. *http://www.nanotechnologies.qc.ca/projets/hydrogene/4/5/2006.*

27. *http://www.abc.net.au/innovations/stories/s1499183.htm 6/29/2008*

28. *http://www.hydrogenpoweredtrucks.com/nasa-press-release 4/8/2006.*

29. *http://www.en.wikipedia.org/wiki/Hydrogen economy 4/5/2006.*

30. *http://www.nmsea.org/Curriculum/7_12/electrolysis. 3/2/2006.*

31. *http://www.eere.energy.gov/hydrogenfuelcells/production. 2/12/2006.*

32. *http://www.ocees.com/textpages/txthydrogen.html 2/19/2006.*

33. *http://www.saharawind.com/HVDCenergytransfer.htm 3/6/2006.*

34. *http://www.nrel.gov/ncpv/documents/25249.html 3/14/2006.*

35. *http://www.thirdworldtraveler.com/oil_watch/Oil_Reserve. 4/15/2006.*

36. *http://www.gas-plants.com/gas-liquefier.html. 2/18/2006.*

37.   http://www.eia.doe.gov/pub/oil_gas/petroleum/analysis. 4/15/2006.
38.   http://www.bellona.no/en/energy/hydrogen/report_6-2002. 4/8/2006.
39.   http://www.culturechange.org/alt_energy.htm. 2/19/2006.
40.   http://www.hydrogennow.org/Facts/Safety-1.htm. 2/7/2006.
41.   http://www.rmi.org/sitepages/pid536.php 2/7/2006.
42.   http://www.veccal.ernet.in/-vecpage/scc/Plant.htm 4/2/2006.
43.   http://www-safety.deas.harvard.edu/services/hydrogen.html
      2/7/2006.
44.   http://www.theglobalist.com/DBWeb/printStoryld.aspx? 4/18/2006.
45.   http://www.nmsea.org/Curriculum/7_12/electrolysis 4/2/2006.
46.   http://www.hydrogenassociation.org/general/faqs.asp 2/17/2006.
47.   http://www.rps.psu.edu/hydrogen/unbound.html 3/15/2006.
48.   http://www.apricus-solar.com/html/insolation_levels_usa
      3/16/2006.
49.   http://www.natureswarriors.org/pipeline.php 3/15/2006.
50.   http://www.veccal.ernet.in/-vecpage/scc/Plant.htm 4/2/2006.
51.   http://www.humbolt.edu/-serc/pemelectrolysis.html 4/16/2006.
52.   http://www.supercables.com/Applications/App_visions.html
      4/25/2006.
53.   http://www.americanenergyindependence.com/hydroge.html
      3/15/2006.
54.   http://www.fromthewilderness.com/free/ww3/081803_hydrogen_
      answers.html 2/20/2006.
55.   http://www.wikipedia.org/wiki/Wind_turbine#Types_of_wind_
      turbines 02/20/2006.
56.   http://www.wikipedia.org/wiki/Kevlar 03/16/2006.
57.   html://www.universalpowercorp.com/hydrogen.htm 03/25/2006.
58.   html://www.hydrogen-carsbiz/hydrogen-fuel-stations.htm
      04/20/2006.
59.   http://www.energybulletin.net/4541html 03/21/2006.
60.   http://www.gkss.de/Themen/w/wtpwassrstaff/eSpeicher.html
      03/25/2006.
61.   http://en.wikipedia.org/wiki/Methane_hydrate 04/06/2006
62.   http://www2.warwick.ac.uk/fac/soc/economics/staff/faculty/oswald/
      windaccountancy04.pdf 02/26/2006.
63.   http://www.tsaugust.org/imiges/hydrogen_reality_and_policy_04_
      update.pdf 03/25/2006.
64.   http://www.storhy.net/pdf/cfc-ICMC_c4-B-02_2003-09-26pdf
      03/19/2006.

65. *http://www.rmi.org/images/other/Energy/E03-05-20HydrogenMyths. pdf 02/26/2006.*

66. *http://www.publicfuelcell.org/sitebuildercontent/sitebuilderfiles/ vehiclesofchange.pdf 04/24/2006.*

67. *http://www.llnl.gov/str/pdfs/03-06pdf 04/05/2006.*

68. *http:/www.linde-kryotechnik.ch/public/facherichte/bulk-hydrogen. pdf 03/08/2006.*

69. *http://www.hydrogennow.org/facts/pipeline/leighty/wec-Sydney-Sept04pdf 03/07/2006.*

70. *http://www.feasta.org/documents/wells/contents.html 04/20/2006.*

71. *http://www.stardrivedevice.com/electrolysis.html 03/10/2006*

72. *http://www.tiretrack.com/tires/tiretech/techpage 11/24/2006*

73. *http://www.gm.com/company/gmability/adv-tech/400_fcv/ hydrogen3_090705html 02/20/2006.*

74. *http://www.bmwi.de/English/redaktion/pdf/strategy-report-on-research-needs-in-the-field-of-hydrogen-energy-technology-daku-546 03/19/2006.*

75. *http://www.minerals.usgs.gov/minerals/pubs/mcs/2006/mcs2006.pdf 03/16/2006.*

76. *http://www.memagazine.org/backissues/feb02/features/fillerup/ fillerup.html 03/10/2006.*

77. *http://www.geography.about.com/cs/censuspopulation/ 02/03/2006.*

78. *http://www.hydrogen.energy.gov/presidents_initiative.html 04/20/2006.*

79. *http://www.eia.doe.gov/pub/oil-gas/petroleum/analysis 04/20/2006.*

80. *http:/www.eia.doe.gov/emeu/steo/pub/a1tab.html 04/20/2006.*

81. *http://www.eia.doe.gov/emeu/steo/pub/5atab.html 04/20/2006.*

82. *http://www.wikipedia.org/wiki/Fischer-Tropsch_process 04/18/2006.*

83. *http;//www.usmm.org/libertyships.html*

84. *http://www.darkendeavors.com/imbedded_images 11/23/2006*

85. *http://www.ece.umr.edu/links/Energy_Course 11/22/2006*

86. *http://www.bergey.com/Maps/World.Wind.Med.htm 11/23/2006*

87. *http:///www.msnbc.msn.com/id/14678206 11/21/2006*

88. *http://www.autobloggreen.com/2006/10/18 11/29/2006*

89. *http://www.nrel.gov/hydrogen/expertise.html 11/27/2006*

90. *http://www.global-hydrogen-bus-platform.com 11/27/2006*

91. *http://www.eere.energy.gov/hydrogenandfuelcells 11/29/2006*

92. *http://www.usinfo.state.gov/gi/archive/2005/apr 11/28/2006*

93. *http://www.popularmechanics.com/blogs 11/27/2006*
94. *http://www.art.com/asp/sp-asp 12/7/2006*
95. *http://www.mnforsustain.org/windpower_cato_part1 2/1/2007*
96. *http:///www.universityof phoenixstadium.com 3/25/08*
97. *http://www.technologyreview.com/19296 2/10/08*
98. *http://www.mhi-ir.jp/20060411104 2/5/08*
99. *http://www.wikipedia.org/wiki/hydroponics 2/11/08*
100. *http://www.technovelgy.com/761 2/10/08*
101. *http://www.popularmechanics.com/1283056 2/10/2008*
102. *http://www.wikipedia.org/wiki/space_habitat 2/12/08*

Index

www.ingramcontent.com/pod-product-compliance
Lightning Source LLC
Chambersburg PA
CBHW031840170526
45157CB00001B/368